Guide for the Total Solar Eclipse of 2045

International Edition

Fred Espenak

Edition 1.0
September 2024

Guide for the Total Solar Eclipse of 2045 – International Edition

Astropixels Publishing
P.O. Box 16197
Portal, AZ 85632

Astropixels Publishing Website: *www.astropixels.com/pubs*

This book may be ordered at: *www.astropixels.com/pubs/Guide2045.html*

Printed in the United States of America

ISBN 978-1-941983-49-2

Astropixels Publication: AP047 (Version 1.0a)

First Edition

Front Cover: Global map of the Aug 12, 2045 total solar eclipse through the USA, West Indies and South America.

Back Cover: Fred Espenak in Mazatlán, Mexico: Copyright ©2024 by Pat Espenak.

More about the eclipse can be found at:

www.*eclipsewise.com/solar/SEnews/TSE2045.html*

Table of Contents

Photo 1 – The Sun's corona is only visible to the naked eye for a few minutes during a total eclipse of the Sun. Total Solar Eclipse of March 29, 2006 from Jalu, Libya. © 2006 F. Espenak, www.MrEclipse.com

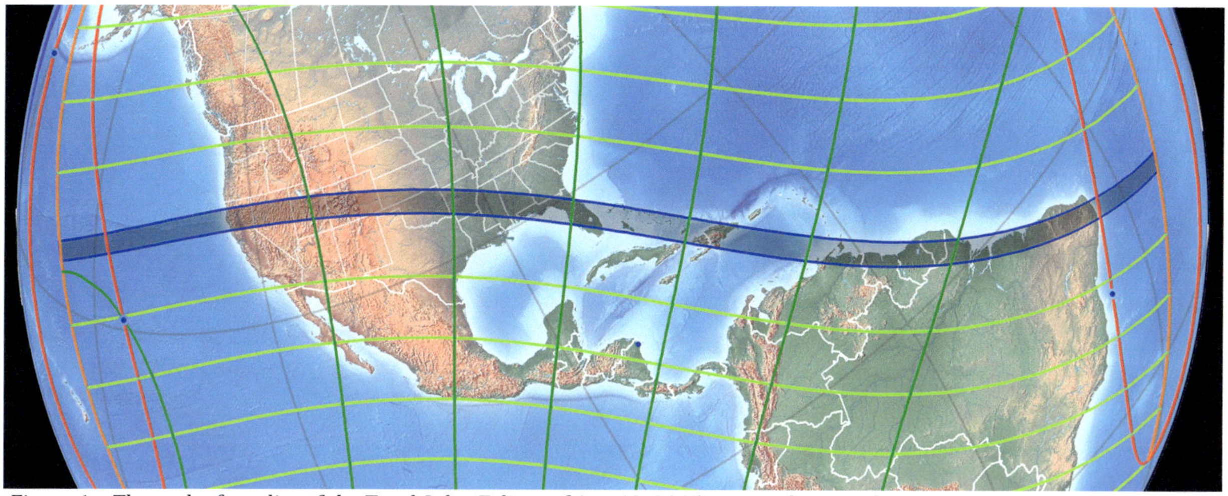

Figure 1 – The path of totality of the Total Solar Eclipse of Aug 12, 2045 crosses the USA, the West Indies, and South America.

1 – Introduction

On Saturday, August 12, 2045, a total eclipse of the Sun will be visible from the Western Hemisphere. Although a partial eclipse will be seen from most of North and South America, the total phase in which the Moon completely covers the Sun (known as totality) will only be seen from within the narrow path of the Moon's umbral shadow as it sweeps across eleven countries and territories: the United States[1], the Bahamas, the Turks and Caicos Islands, Haiti, Dominican Republic, Trinidad and Tobago, Venezuela, Guyana, Suriname, French Guiana, and Brazil.

During totality, the Sun's faint and exquisitely beautiful outer atmosphere – the solar corona – can be seen with the naked eye, the landscape is plunged into an eerie twilight, and brighter stars and planets appear. A total solar eclipse is one of the most awesome natural spectacles, but to witness it you must be in the *path of totality*.

The path of the Moon's umbral shadow begins in the North Pacific and travels east to reach the coast of California. The central duration of totality already exceeds four minutes as the path continues across Nevada and Utah. In Colorado, the duration is over five minutes. Onwards through Kansas, Oklahoma, Arkansas, and Mississippi, the duration surpasses the six-minute mark in Alabama and Florida.

Leaving the USA, the eclipse track crosses the West Indies including the Bahamas, Haiti, and the Dominican Republic. It reaches the Caribbean coast of Venezuela and hugs the northern coasts of Guyana and Suriname. After crossing French Guiana, the umbral shadow path sweeps through northern Brazil. The last three minutes of the total eclipse track brings the Moon's umbra across the Atlantic Ocean where it ends at sunset.

A partial eclipse will be visible from a much larger region including most of North and South America. Eclipse glasses or some type of eye protection must be used for viewing the partial eclipse. Of course, those in the path of the total eclipse must remove protective glasses to witness totality itself.

This guide is designed to assist observers in locating and reaching the central eclipse path to enjoy the awesome experience of totality.

[1] The path of the total solar eclipse of Aug 12, 2045 crosses parts of fifteen states: California, Nevada, Utah, Colorado, Kansas, Texas, New Mexico, Oklahoma, Missouri, Arkansas, Mississippi, Louisiana, Alabama, Georgia, and Florida. This is the first total eclipse to cross the USA from coast-to-coast (Pacific to Atlantic) since Aug 21, 2017 (14 states). The total eclipse of Apr 8, 2024 crossed the USA from Texas to Maine (15 states) plus Mexico and Canada. The upcoming total eclipse of Mar 30, 2033 crosses northern Alaska and Russia, while the eclipse of Aug 23, 2044 crosses North Dakota and Montana (plus Canada and Greenland).

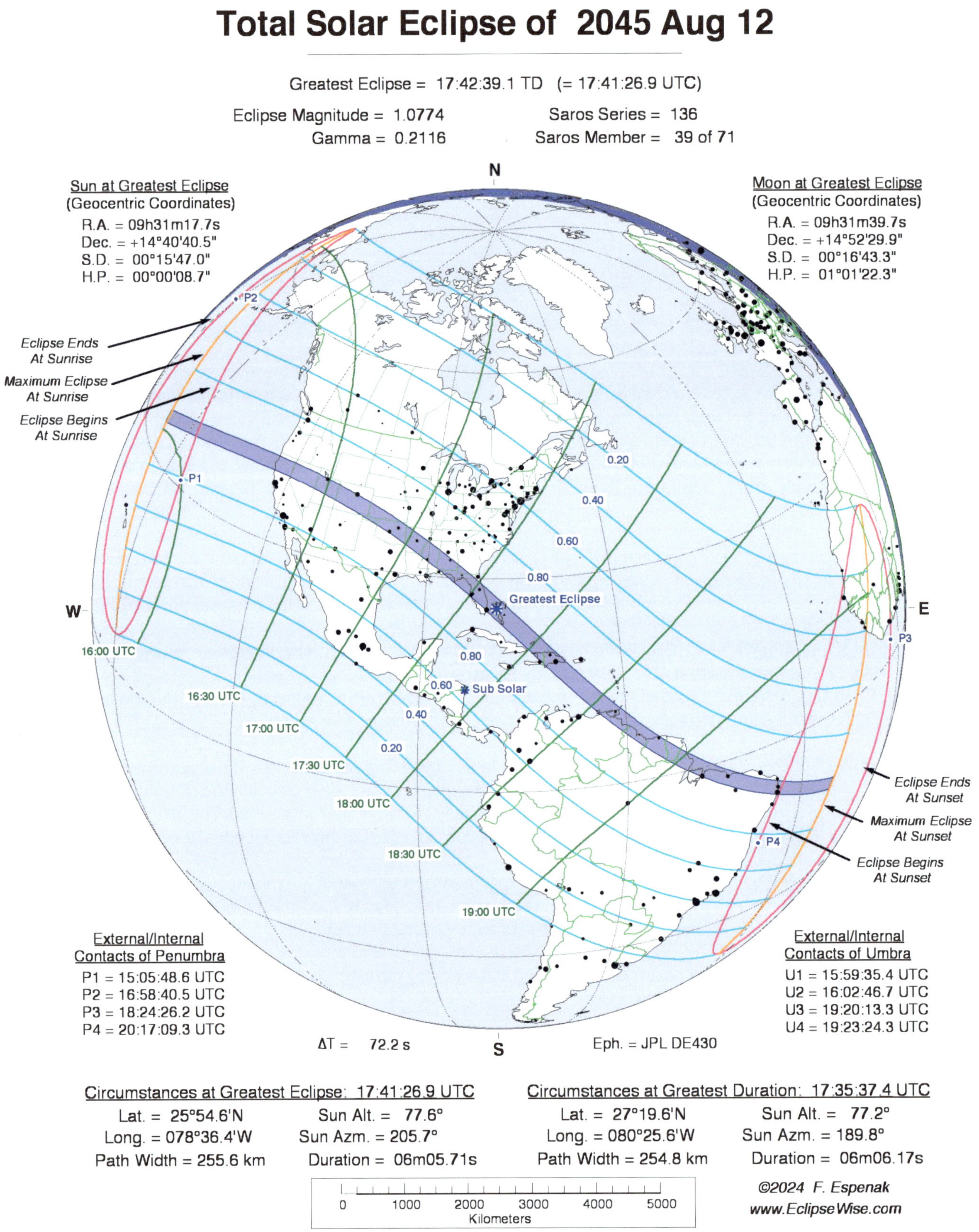

Figure 2 – Orthographic projection map of the Total Solar Eclipse of 2045.
Based on map from ***21st Century Canon of Solar Eclipses*** *by Fred Espenak (2017)*

2 – Orthographic Projection Map of the 2045 Total Solar Eclipse Path

The orthographic projection map of Earth in *Figure 2* shows the path of penumbral (partial) and umbral (total) shadows of the Moon during the eclipse.

The limits of the Moon's penumbral shadow define the region of visibility of the partial eclipse. It covers about half of Earth's daylight hemisphere and consists of several distinct zones or limits. At the northern and southern boundaries lie the limits of the penumbra's path. Great loops at the western and eastern extremes of the penumbra's path identify the areas where the eclipse begins and ends at sunrise and sunset, respectively.

Bisecting the *eclipse begins and ends at sunrise and sunset* loops is the curve of maximum eclipse at sunrise (western loop) and sunset (eastern loop). Points *P1* and *P4* mark the coordinates where the penumbral shadow first contacts (partial eclipse begins) and last contacts (partial eclipse ends) Earth's surface. The path of the umbral shadow (*path of totality*) appears within the penumbral path. Curves of constant eclipse magnitude define points where the maximum magnitude has values of 0.2, 0.4, 0.6, and 0.8.

Above the map is the instant of greatest eclipse[2], in Terrestrial Time[3] (TT) and Coordinated Universal Time[4] (UTC). The *Eclipse Magnitude* is the geocentric ratio of diameters of the Moon and the Sun. *Gamma* is the minimum distance of the Moon's shadow axis from Earth's center in Earth radii at greatest eclipse. Finally, the *Saros Series* of the eclipse is listed, followed by the *Saros Member* that gives the sequence position of the eclipse in the Saros, and the total number of eclipses in the series

In the upper left and right corners are the geocentric coordinates of the Sun and the Moon, respectively, at the instant of greatest eclipse. They are:

- R.A. – Right Ascension
- Dec. – Declination
- S.D. – Apparent Semi-Diameter
- H.P. – Horizontal Parallax

To the lower left are the contact times of the Moon's penumbral shadow with Earth. They mark the times when the partial eclipse begins and ends, globally. To the lower right are the contact times of the Moon's umbral shadow with Earth. They mark the times when the total eclipse path begins and ends, globally.

At bottom left are the geographic coordinates of the position of greatest eclipse along with the local circumstances at that location (i.e., Sun altitude, Sun azimuth, path width and duration of totality).

At bottom right are the corresponding values for the position of greatest duration[5]. Although the geographic locations of greatest duration and greatest eclipse differ by about 240 km (149 miles) along the eclipse track, the difference in the duration of totality is only 0.46 second.

Photo 2 – Time sequence of the Total Solar Eclipse of August 21, 2017, from Casper, Wyoming.
This sequence shows the partial phases, the diamond ring effect, and the solar corona during totality.
© 2017 F. Espenak, www.MrEclipse.com

---

[2] The instant of greatest eclipse occurs when the Moon's shadow axis passes closest to Earth's center.

[3] Terrestrial Time or TT is a time scale established by astronomers to calculate the location or planetary position of other objects as they relate to the center of Earth and its rotation.

[4] Coordinated Universal Time or UTC is the primary time standard that regulates world clocks and time. It is based on International Atomic Time (TAI) with leap seconds added at irregular intervals to compensate for the slowing of Earth's rotation. UTC is not adjusted for daylight saving time. It is effectively a successor to Greenwich Mean Time (GMT).

[5] Greatest duration is the time (and location) when the length of totality reaches its maximum value.

Figure 3 – Path of the Total Solar Eclipse of Aug 12, 2045 through the USA.

3 – Path of Totality

The path of the Moon's umbral shadow begins at 16:01 UTC in the North Pacific about 1590 km (990 mi) north of the Hawaiian Islands. Along the sunrise terminator, the duration of totality is 3m 1s (i.e., 3 minutes 1 seconds) as seen from the center of the narrow, 204 km (127 mi) wide path. No landfall occurs during the first 15 minutes of the central eclipse as the shadow sweeps 3140 km (1950 mi) northeast across the Pacific.

The path of the total eclipse across the USA begins at 16:16 UTC (9:16 am PDT) along the Pacific coast of northern California. The central duration of totality is 4m 23s, the Sun's altitude is 31° and the track width is 228 km (142 mi). The largest California city in the path is Redding (pop. 93,000) where its citizens will witness 4m 25s of totality.

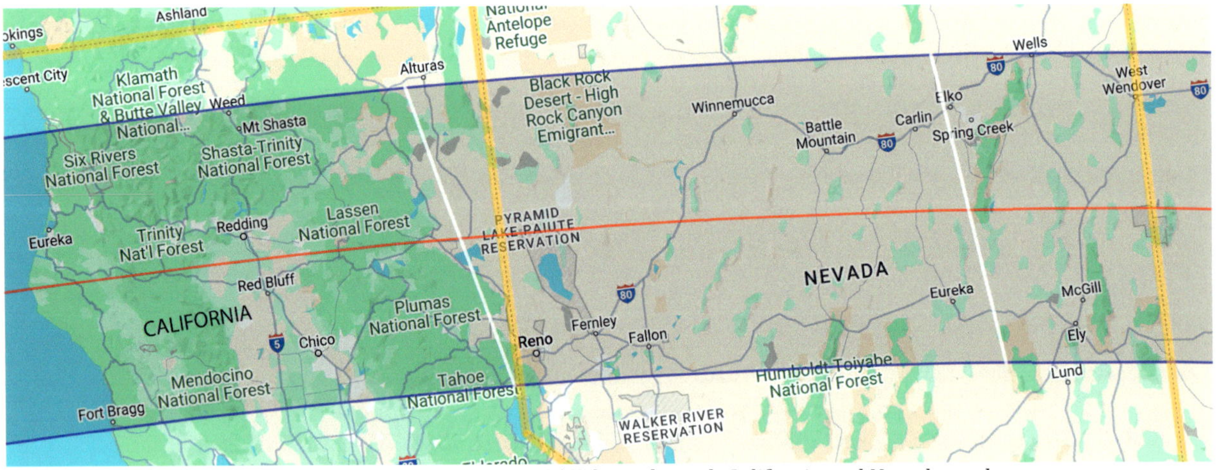

Figure 4 – Path of the Aug 12, 2045 Eclipse through California and Nevada, and.

Continuing over the Sierra Nevada Range, the track enters Nevada. Lady luck shines of Reno (pop. 243,000) which lies within the path. Although 93 km (58 mi.) south of the central line, Reno's population still enjoys 2m 45s of totality. Las Vegas, Carson City, and Lake Tahoe do not fare as well because they all lie outside the path and only witness a deep partial eclipse.

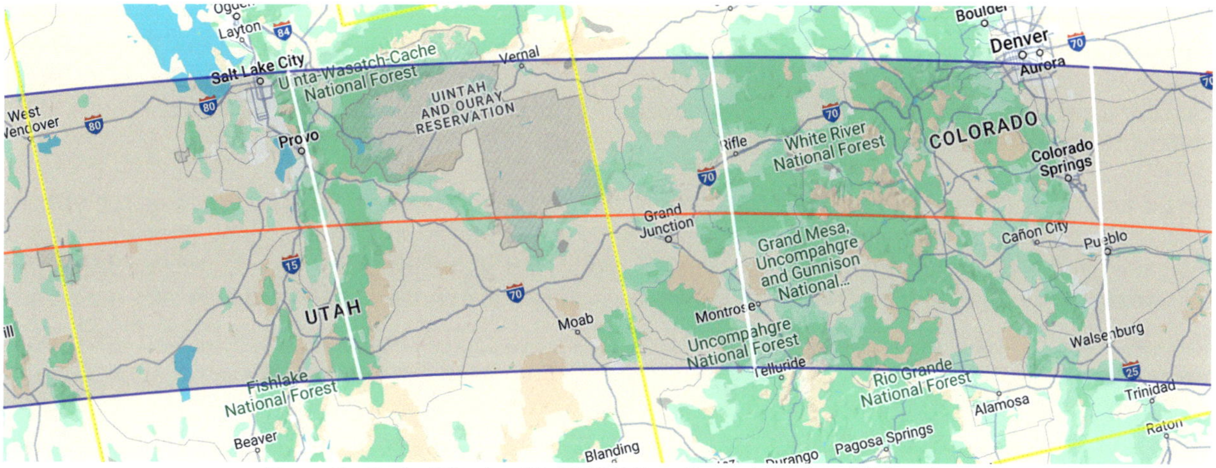

Figure 5 – Path of the Aug 12, 2045 Eclipse through Utah and Colorado.

Racing across Utah, the state's two largest cities Salt Lake City and Provo lie within the umbra's path. As it exits the "Beehive State," the path begins to curve to the southeast and the central duration exceeds five minutes. Colorado Springs is the next large city (pop. 486,000) in the path with a duration of 5m 4s and a Sun altitude of 50°. Denver, Colorado's capital, is just outside the northern path limit although some southern regions within the metropolitan area will witness a brief totality.

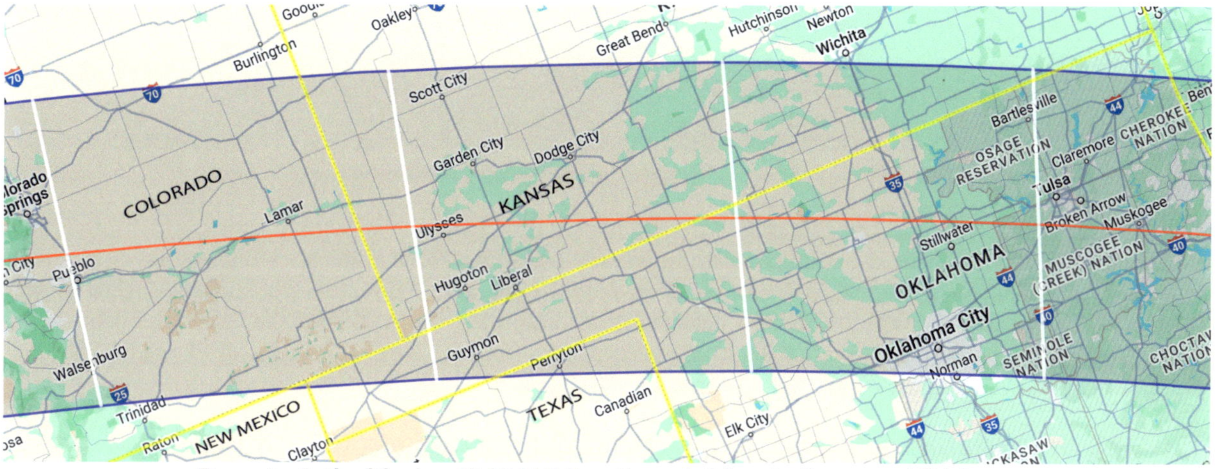

Figure 6 – Path of the Aug 12, 2045 Eclipse through Colorado, Kansas, and Oklahoma.

A tiny corner (approximately 40 sq. km) of northeastern New Mexico is grazed by the southern edge of the path as it enters the panhandle of Oklahoma. The path also includes southwestern Kansas and northeastern corner of Texas' panhandle. In Oklahoma, Oklahoma City (pop. 695,000) is the largest city in the path with a duration of 3m 26s, but smaller Tulsa (pop. 412,000) is closer to center where totality lasts 5m 33s.

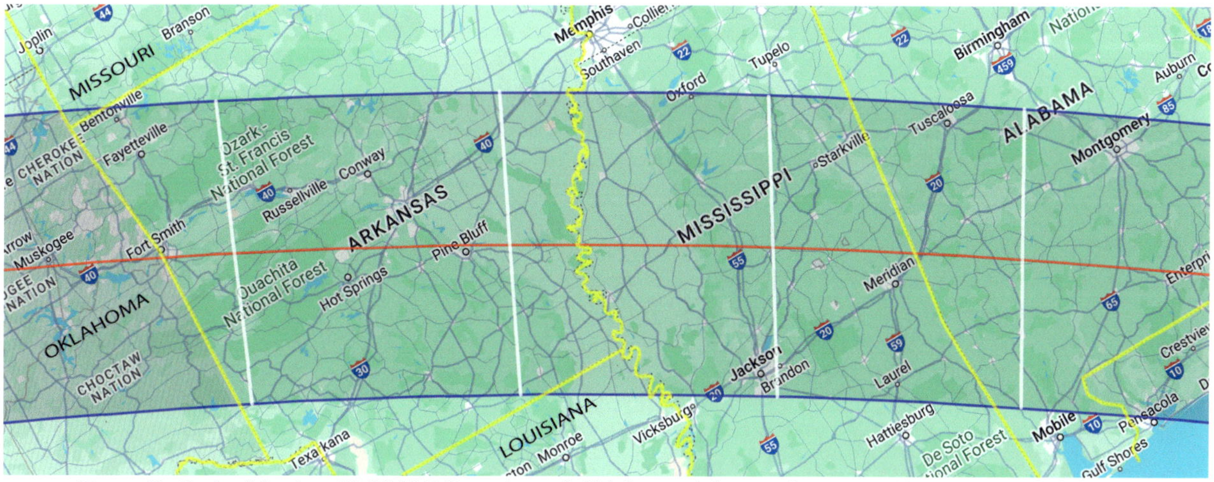

Figure 7 – Path of the Aug 12, 2045 Eclipse through Oklahoma, Arkansas, Louisiana, Mississippi, and Alabama.

The southwestern corner (approximately 230 sq. km) of Missouri is clipped by the path as it crosses into Arkansas. Just over the border, the inhabitants of Fort Smith (pop. 90,000) get totality lasting 5m 44s with the Sun 62° high. But the largest city in the Arkansas path, also its state capital, is Little Rock (pop. 203,000) which witnesses a total eclipse lasting 5m 36s.

As the umbra's track traverses the Mississippi River, it enters the northeast corner of Louisiana and cuts across Mississippi where the state capital, Jackson (pop. 146,000), receives 3m 10s of totality. Proceeding on to Alabama, Montgomery (pop. 197,000), another state capital, lies within the path with a total eclipse lasting 3m 40s. Spanning the "Cotton State," the track crosses the southwestern corner of Georgia before entering Florida.

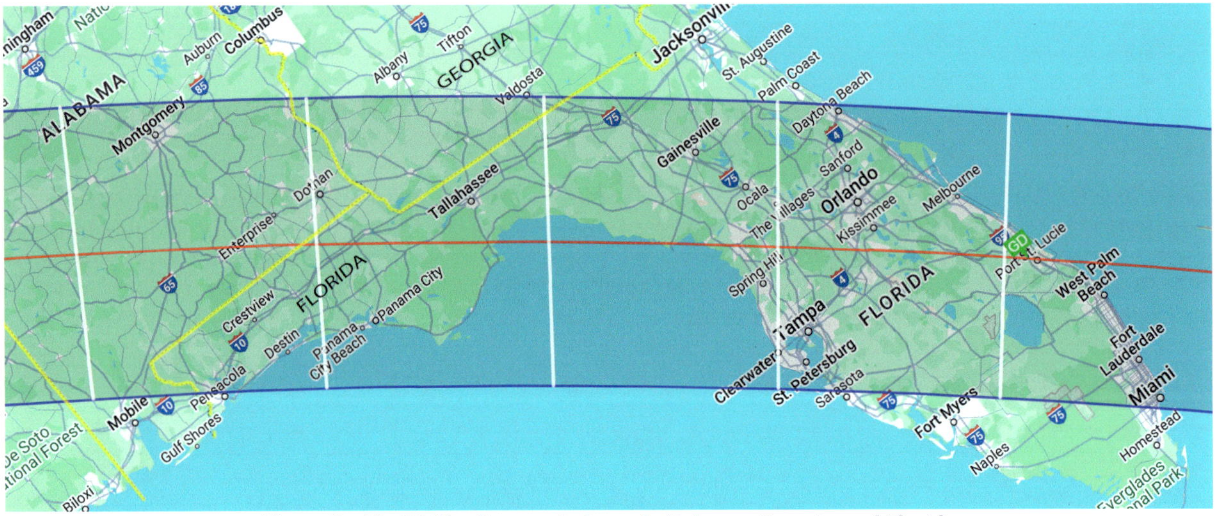

Figure 8 – The 2045 eclipse path through Alabama, Georgia, and Florida.

The "Sunshine State" is the most populated section of the eclipse track. Sixteen cities with populations exceeding 100,000 are located within the path of totality including Fort Lauderdale, Gainesville, Orlando, St. Petersburg, Tallahassee (state capital), Tampa, West Palm Beach, and Miami. It is not surprising that most of Florida's largest cities are within the path because it encompasses about 80% of the state. The eclipse also reaches its maximum with the point of *Greatest Duration* occurring at 17:36:27 UTC (1:36:27 pm EDT) in Port St. Lucie. The duration on the central line is 6m 6.2s.

The path of totality through the USA crosses parts of fifteen states (California, Nevada, Utah, Colorado, Kansas, Texas, New Mexico, Oklahoma, Missouri, Arkansas, Mississippi, Louisiana, Alabama, Georgia, and Florida). It is 4360 km (2710 mi) long and the Moon's umbral shadow traverses it in 81 minutes – an average velocity of 3230 km/hr (2007 mi/hr).

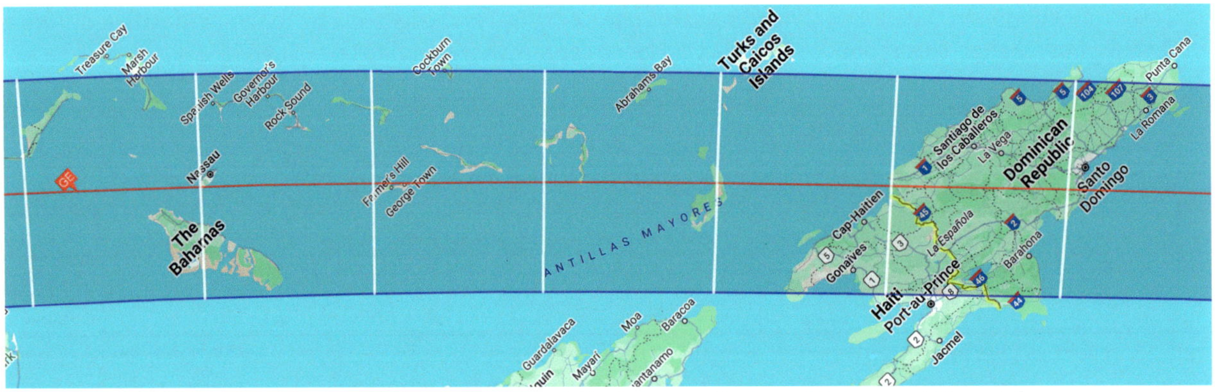

Figure 9 – The 2045 eclipse path through the Bahamas, Turks and Caicos, and Hispaniola.

After leaving the Atlantic coast of Florida, the shadow sweeps over the Lucayan Archipelago consisting of the Bahamas and the Turks and Caicos Islands. The point of *Greatest Eclipse* (the shadow axis passes closest to the center of Earth) occurs at sea at 17:41:27 UTC (1:41:27 EDT) about 66 km (41 mi) south of Freeport, Grand Bahama. The duration of totality is 6m 5.7s and the Sun's altitude is 78°.

Hispaniola is the most populous island in the West Indies. Its two nations (Haiti and the Dominican Republic) are both within the path of totality. Santo Domingo, the capital of the Dominican Republic, is one of the Caribbean's oldest cities. Its 4,275,000 residents will experience a total eclipse lasting 5m 44s at 18:10 UTC (2:10 pm AST).

Crossing the eastern Caribbean Sea, Isla La Blanquilla and Isla de Margarita are in the path just before the shadow reaches the northern coastline of Venezuela. While the entire island of Trinidad is inside the path, the island of Tobago is outside.

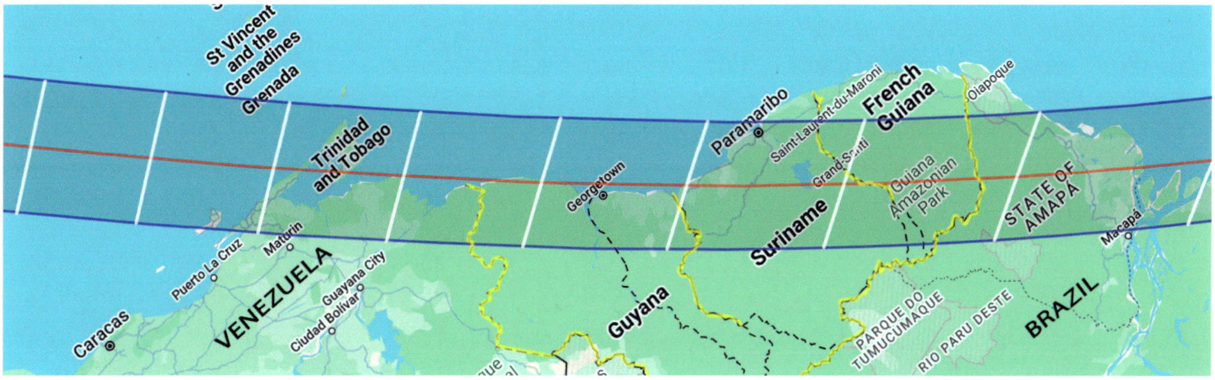

Figure 10 – Path of the Aug 12, 2045 Eclipse through Venezuela, Guyana, Suriname, and French Guiana.

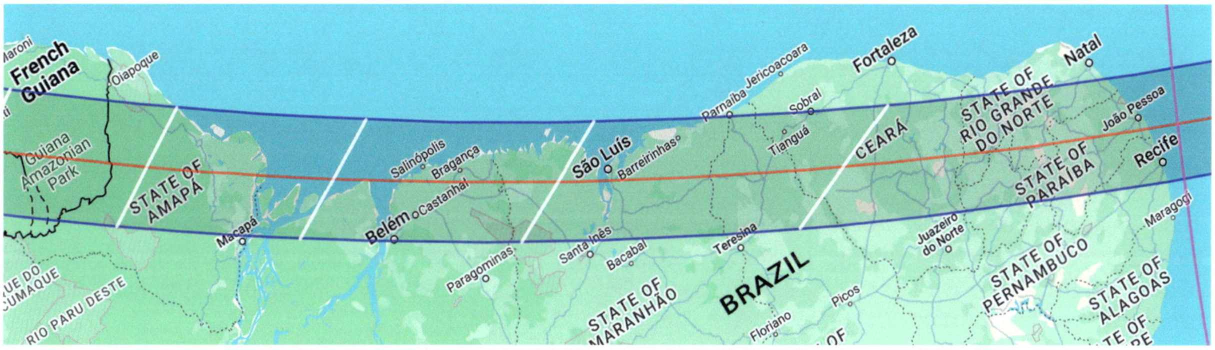

Figure 11 – Path of the Aug 12, 2045 Eclipse through French Guiana and Brazil.

The eclipse track remains close to the Atlantic coast of South America as it crosses Guyana, Suriname, French Guiana, and Brazil. São Luís is the capital and largest city of the Brazilian state of Maranhão. Its population of 1,536,000 residents will enjoy a totality lasting 3m 59s with the late afternoon Sun 26° above the horizon.

Leaving Brazil at 19:19 UTC (4:18 pm BRT), the umbra races 1460 km (905 mi) across the Atlantic before reaching the end of the track at 19:22 UTC and returning to space. Over a period of 3 hours 21 minutes, the umbra travels along a 15,199 km (9444 mi) track that covers an area equal to 0.71% of Earth's surface.

Table 1 – Time Zones used by the USA and Countries within the 2045 Path of Totality

Time Zone Name	Abbrev.	UTC Offset	Regions in 2045 Path
Pacific Daylight Time	PDT	UTC - 7 hrs	CA, NV
Mountain Daylight Time	MDT	UTC - 6 hrs	UT, CO, NM
Central Daylight Time	CDT	UTC - 5 hrs	KS, OK, TX, AR, LA, MS, AL, west FL
Eastern Daylight Time	EDT	UTC - 4 hrs	GE, FL, Bahamas, Hati
Atlantic Standard Time	AST	UTC - 4 hrs	Dominican Republic, Trinidad
Venezuela Standard Time	VST	UTC - 4 hrs	Venezuela
Guyana Time	GYT	UTC - 4 hrs	Guyana
Suriname Time	SRT	UTC - 3 hrs	Suriname
Brasilia Time	BRT	UTC - 3 hrs	Brazil

Table 2 – The 2045 Path of Totality

UTC	Latitude	Longitude	Diameter Ratio	Sun Alt	Sun Azm	Path Width km	Major Axis km	Minor Axis km	Duration of Totality
16:01.2	35°07.7'N	159°41.0'W	1.059	0.0	71.9	204.2	—	193.5	03m01.5s
16:10	40°02.6'N	132°35.7'W	1.066	23.6	90.3	223.8	542.4	216.7	04m01.7s
16:20	40°22.8'N	120°26.1'W	1.070	34.7	100.6	231.6	398.3	226.6	04m34.0s
16:30	39°41.7'N	111°38.9'W	1.072	43.2	109.1	237.1	340.8	233.3	04m58.8s
16:40	38°28.4'N	104°39.4'W	1.073	50.5	117.0	241.5	309.0	238.3	05m18.8s
16:50	36°54.4'N	098°50.8'W	1.075	56.9	124.9	245.0	289.0	242.1	05m34.9s
17:00	35°05.5'N	093°53.2'W	1.076	62.7	133.4	247.9	275.7	245.0	05m47.4s
17:10	33°05.4'N	089°33.8'W	1.076	68.0	143.6	250.3	266.7	247.2	05m56.6s
17:20	30°56.3'N	085°43.5'W	1.077	72.5	157.0	252.3	260.8	248.8	06m02.7s
17:30	28°39.6'N	082°15.3'W	1.077	76.0	176.0	254.0	257.4	249.7	06m05.7s
17:40	26°15.9'N	079°03.2'W	1.077	77.6	201.7	255.4	256.1	250.1	06m05.9s
17:50	23°45.8'N	076°02.3'W	1.077	76.8	228.5	256.7	256.7	249.9	06m03.4s
18:00	21°09.2'N	073°07.9'W	1.077	73.8	249.0	257.7	259.5	249.1	05m58.2s
18:10	18°25.8'N	070°15.1'W	1.077	69.5	262.6	258.4	264.5	247.7	05m50.6s
18:20	15°34.8'N	067°18.9'W	1.076	64.5	271.5	258.7	272.4	245.7	05m40.5s
18:30	12°35.0'N	064°12.9'W	1.075	58.8	277.6	258.3	284.2	243.0	05m28.1s
18:40	09°24.2'N	060°48.5'W	1.074	52.6	281.9	256.9	301.7	239.5	05m13.0s
18:50	05°59.0'N	056°52.8'W	1.072	45.6	284.9	253.9	328.9	234.8	04m55.2s
19:00	02°12.6'N	052°02.0'W	1.070	37.5	286.8	248.4	375.9	228.7	04m33.8s
19:10	02°10.8'S	045°20.3'W	1.067	27.4	287.8	238.7	478.2	220.1	04m07.0s
19:20	08°26.8'S	032°11.3'W	1.062	10.7	286.8	218.0	1101.4	204.1	03m25.1s
19:21.8	11°35.8'S	022°17.7'W	1.059	0.0	285.0	203.7	—	193.3	03m01.3s

Table 2 lists the geographic coordinates of the central line for every ten minutes UTC. *Diameter Ratio* gives the Moon's apparent diameter divided by the Sun's. The Sun's altitude (*Alt*) and azimuth (*Azm*) are given along with the width of the path of totality (kilometers). The *major* and *minor* axes of the umbral shadow are also listed. Finally, the *Duration of Totality* is given (minutes and seconds).

4 – Local Circumstances for the Eclipse

Local circumstances for the ***Total Eclipse*** for 162 cities are found in Table 3 (USA – Alabama to Kansas), Table 4 (USA - Florida), Table 5 (USA – Mississippi to Utah), and Table 6 (West Indies and South America). All times are given in local time (see Table 1 for Time Zones). The *Duration of Totality* is the period between the beginning and end of totality. These times include the effects of the Moon's limb profile. The Sun's altitude is given at the instant of mid-totality.

Local circumstances for the ***Partial Eclipse*** for 102 cities are found in Table 7 (USA) and Table 8 (North, Central, and South America). All times are given in local time and include *Daylight Saving Time*. The Sun's altitude is given at the instant of maximum eclipse. *Eclipse Magnitude* is given as the fraction of the Sun's *diameter* covered by the Moon at maximum eclipse.

The 2045 Total Solar Eclipse Circumstances Calculator is an interactive web page that can quickly calculate the local circumstances for the solar eclipse from any geographic location not included in Tables 3 through 8. The URL address for the calculator is:

https://www.eclipsewise.com/solar/SEcirc/2001-2100/SE2045Aug12Tcirc.html

Table 3 – Total Eclipse Circumstances for Alabama to Kansas

Location	Partial Eclipse Begins	Total Eclipse Begins	Total Eclipse Ends	Partial Eclipse Ends	Duration of Totality	Sun Altitude
Alabama						
Dothan	10:59:06 am	12:17:21 pm	12:23:05 pm	01:40:23 pm	05m40s	72°
Enterprise	10:58:12 am	12:16:16 pm	12:22:11 pm	01:39:29 pm	05m53s	72°
Montgomery	10:56:29 am	12:15:03 pm	12:18:43 pm	01:36:51 pm	03m40s	71°
Northport	10:53:32 am	12:11:59 pm	12:14:46 pm	01:33:14 pm	02m47s	69°
Prattville	10:56:07 am	12:14:37 pm	12:18:20 pm	01:36:26 pm	03m44s	71°
Selma	10:55:13 am	12:13:00 pm	12:18:05 pm	01:35:40 pm	05m06s	70°
Tuscaloosa	10:53:41 am	12:12:07 pm	12:15:00 pm	01:33:25 pm	02m54s	69°
Arkansas						
Benton	10:44:35 am	11:59:56 am	12:05:45 pm	01:22:46 pm	05m50s	64°
Conway	10:44:35 am	12:00:03 pm	12:05:08 pm	01:22:16 pm	05m05s	64°
Fayetteville	10:41:35 am	11:56:34 am	12:00:37 pm	01:17:52 pm	04m09s	62°
Florence	10:46:29 am	12:02:30 pm	12:08:09 pm	01:25:36 pm	05m41s	66°
Fort Smith	10:41:25 am	11:55:53 am	12:01:38 pm	01:18:21 pm	05m44s	62°
Hot Springs	10:43:53 am	11:59:12 am	12:04:52 pm	01:22:03 pm	05m44s	64°
Jacksonville	10:45:11 am	12:00:43 pm	12:06:03 pm	01:23:08 pm	05m16s	64°
Japton	10:42:10 am	11:57:24 am	12:01:15 pm	01:18:38 pm	03m56s	62°
Little Rock	10:44:58 am	12:00:22 pm	12:06:01 pm	01:23:01 pm	05m36s	64°
Pine Bluff	10:45:39 am	12:01:19 pm	12:07:08 pm	01:24:17 pm	05m51s	65°
Rogers	10:41:35 am	11:57:20 am	11:59:38 am	01:17:34 pm	02m14s	62°
Russellville	10:43:25 am	11:58:29 am	12:03:51 pm	01:20:46 pm	05m18s	63°
Springdale	10:41:36 am	11:56:51 am	12:00:17 pm	01:17:45 pm	03m29s	62°
California						
Chico	08:13:52 am	09:16:40 am	09:20:05 am	10:28:46 am	03m25s	33°
Eureka	08:13:17 am	09:14:36 am	09:18:44 am	10:25:41 am	04m07s	31°
Paradise	08:14:00 am	09:16:49 am	09:20:23 am	10:29:04 am	03m34s	34°
Red Bluff	08:13:52 am	09:15:59 am	09:20:18 am	10:28:12 am	04m22s	33°
Redding	08:14:01 am	09:15:56 am	09:20:21 am	10:28:00 am	04m25s	33°
Susanville	08:14:52 am	09:17:30 am	09:22:02 am	10:30:25 am	04m33s	35°
Truckee	08:14:37 am	-	09:20:10 am	10:31:07 am	00m00s	35°
Colorado						
Canon City	09:26:31 am	10:35:54 am	10:40:59 am	11:55:06 am	05m09s	49°
Colorado Springs	09:27:30 am	10:37:04 am	10:42:11 am	11:56:16 am	05m04s	50°
Englewood	09:27:31 am	10:38:35 am	10:39:45 am	11:55:19 am	01m04s	50°
Grand Junction	09:23:40 am	10:31:39 am	10:36:40 am	11:49:45 am	05m04s	46°
Littleton	09:27:29 am	10:38:18 am	10:40:02 am	11:55:19 am	01m41s	50°
Parker	09:27:43 am	10:38:23 am	10:40:40 am	11:55:47 am	02m16s	50°
Pueblo	09:27:40 am	10:37:32 am	10:42:41 am	11:57:09 am	05m13s	51°
Georgia						
Bainbridge	12:00:53 pm	01:19:30 pm	01:24:46 pm	02:42:14 pm	05m16s	73°
Thomasville	12:02:04 pm	01:21:15 pm	01:25:29 pm	02:43:19 pm	04m19s	73°
Kansas						
Derby	10:36:51 am	11:51:31 am	11:52:55 am	01:10:35 pm	01m19s	58°
Dodge City	10:33:10 am	11:45:03 am	11:50:06 am	01:05:37 pm	05m01s	55°
Garden City	10:32:05 am	11:43:32 am	11:48:37 am	01:03:53 pm	05m03s	54°
Liberal	10:32:04 am	11:43:56 am	11:48:58 am	01:04:47 pm	05m01s	55°

Table 4 – Total Eclipse Circumstances for Florida

Location	Partial Eclipse Begins	Total Eclipse Begins	Total Eclipse Ends	Partial Eclipse Ends	Duration of Totality	Sun Altitude
Boca Raton	12:14:47 pm	01:35:13 pm	01:40:34 pm	02:57:39 pm	05m23s	78°
Boynton Beach	12:14:34 pm	01:34:44 pm	01:40:27 pm	02:57:20 pm	05m45s	78°
Bradenton	12:08:27 pm	01:30:16 pm	01:32:22 pm	02:51:39 pm	02m04s	77°
Clearwater	12:07:25 pm	01:28:00 pm	01:32:08 pm	02:50:26 pm	04m09s	77°
Coral Gables	12:15:17 pm	01:37:47 pm	01:39:33 pm	02:58:29 pm	01m39s	79°
Coral Springs	12:14:40 pm	01:35:15 pm	01:40:21 pm	02:57:36 pm	05m05s	78°
Davie	12:14:53 pm	01:36:04 pm	01:40:12 pm	02:57:57 pm	04m09s	78°
Daytona Beach	12:09:24 pm	01:30:40 pm	01:32:09 pm	02:50:49 pm	01m22s	75°
De Land	12:09:03 pm	01:29:03 pm	01:33:14 pm	02:50:46 pm	04m15s	76°
Deerfield Beach	12:14:48 pm	01:35:18 pm	01:40:33 pm	02:57:41 pm	05m17s	78°
Delray Beach	12:14:40 pm	01:34:54 pm	01:40:30 pm	02:57:27 pm	05m38s	78°
Deltona	12:09:20 pm	01:29:12 pm	01:33:43 pm	02:51:07 pm	04m35s	76°
Fort Lauderdale	12:15:01 pm	01:35:51 pm	01:40:35 pm	02:58:01 pm	04m39s	78°
Fort Pierce	12:12:50 pm	01:32:34 pm	01:38:36 pm	02:55:12 pm	06m03s	77°
Fort Walton Beach	10:57:40 am	12:17:05 pm	12:21:01 pm	01:39:48 pm	03m57s	73°
Gainesville	12:06:22 pm	01:25:56 pm	01:30:28 pm	02:48:01 pm	04m36s	75°
Hialeah	12:15:07 pm	01:37:15 pm	01:39:41 pm	02:58:19 pm	02m27s	78°
Hollywood	12:15:08 pm	01:36:17 pm	01:40:28 pm	02:58:10 pm	04m12s	78°
Kissimmee	12:09:42 pm	01:29:11 pm	01:35:11 pm	02:52:00 pm	06m00s	76°
Lakeland	12:08:56 pm	01:28:34 pm	01:34:26 pm	02:51:37 pm	05m53s	77°
Largo	12:07:30 pm	01:28:13 pm	01:32:09 pm	02:50:33 pm	03m58s	77°
Lauderhill	12:14:50 pm	01:35:47 pm	01:40:18 pm	02:57:51 pm	04m28s	78°
Margate	12:14:42 pm	01:35:23 pm	01:40:21 pm	02:57:40 pm	04m54s	78°
Melbourne	12:11:30 pm	01:31:11 pm	01:36:46 pm	02:53:35 pm	05m33s	77°
Miami	12:15:24 pm	01:37:26 pm	01:40:04 pm	02:58:33 pm	02m42s	78°
Miami Beach	12:15:29 pm	01:37:24 pm	01:40:18 pm	02:58:38 pm	03m00s	78°
Miramar	12:14:59 pm	01:36:28 pm	01:40:03 pm	02:58:05 pm	03m41s	78°
North Miami	12:15:14 pm	01:36:51 pm	01:40:14 pm	02:58:20 pm	03m30s	78°
Ocala	12:07:16 pm	01:26:35 pm	01:32:08 pm	02:49:14 pm	05m31s	75°
Orlando	12:09:26 pm	01:28:55 pm	01:34:40 pm	02:51:33 pm	05m42s	76°
Palm Bay	12:11:37 pm	01:31:18 pm	01:36:56 pm	02:53:43 pm	05m36s	77°
Panama City	10:59:36 am	12:18:37 pm	12:23:46 pm	01:41:48 pm	05m10s	73°
Pembroke Pines	12:14:59 pm	01:36:22 pm	01:40:09 pm	02:58:05 pm	03m50s	78°
Pinellas Park	12:07:44 pm	01:28:35 pm	01:32:18 pm	02:50:48 pm	03m47s	77°
Plantation	12:14:46 pm	01:35:47 pm	01:40:11 pm	02:57:48 pm	04m22s	78°
Pompano Beach	12:14:51 pm	01:35:30 pm	01:40:32 pm	02:57:48 pm	05m00s	78°
Port Orange	12:09:32 pm	01:30:32 pm	01:32:35 pm	02:51:00 pm	01m58s	76°
Port St. Lucie	12:12:57 pm	01:32:43 pm	01:38:48 pm	02:55:24 pm	06m06s	77°
St. Petersburg	12:07:54 pm	01:28:49 pm	01:32:27 pm	02:50:59 pm	03m42s	77°
Sunrise	12:14:56 pm	01:36:00 pm	01:40:20 pm	02:57:58 pm	04m19s	78°
Tallahassee	12:01:53 pm	01:20:29 pm	01:26:20 pm	02:43:33 pm	05m47s	74°
Tamarac	12:14:48 pm	01:35:36 pm	01:40:22 pm	02:57:47 pm	04m42s	78°
Tampa	12:08:06 pm	01:28:14 pm	01:33:16 pm	02:51:01 pm	05m00s	77°
Titusville	12:10:28 pm	01:30:28 pm	01:34:55 pm	02:52:16 pm	04m31s	76°
West Palm Beach	12:14:21 pm	01:34:21 pm	01:40:18 pm	02:57:01 pm	05m58s	78°
Winter Haven	12:09:24 pm	01:28:59 pm	01:34:59 pm	02:52:01 pm	06m00s	77°

Table 5 – Total Eclipse Circumstances for Mississippi to Utah

Location	Partial Eclipse Begins	Total Eclipse Begins	Total Eclipse Ends	Partial Eclipse Ends	Duration of Totality	Sun Altitude
Mississippi						
Brandon	10:50:12 am	12:08:22 pm	12:12:00 pm	01:30:52 pm	03m41s	68°
Clarksdale	10:47:59 am	12:04:13 pm	12:09:28 pm	01:26:46 pm	05m13s	66°
Cleveland	10:48:00 am	12:04:11 pm	12:10:03 pm	01:27:15 pm	05m53s	67°
Clinton	10:49:34 am	12:08:06 pm	12:10:53 pm	01:30:08 pm	02m54s	68°
Columbus	10:52:01 am	12:09:41 pm	12:13:30 pm	01:31:31 pm	03m52s	68°
Greenville	10:47:38 am	12:04:02 pm	12:09:34 pm	01:27:11 pm	05m34s	67°
Greenwood	10:49:02 am	12:05:25 pm	12:11:18 pm	01:28:32 pm	05m54s	67°
Jackson	10:49:49 am	12:08:14 pm	12:11:19 pm	01:30:25 pm	03m10s	68°
Laurel	10:52:06 am	12:11:15 pm	12:13:54 pm	01:33:18 pm	02m45s	70°
Meridian	10:52:19 am	12:09:32 pm	12:15:22 pm	01:32:56 pm	05m51s	69°
Pearl	10:49:55 am	12:08:24 pm	12:11:25 pm	01:30:34 pm	03m07s	68°
Ridgeland	10:49:50 am	12:07:35 pm	12:11:46 pm	01:30:20 pm	04m08s	68°
Starkville	10:51:21 am	12:08:30 pm	12:13:18 pm	01:30:55 pm	04m51s	68°
Nevada						
Elko	08:18:14 am	09:23:21 am	09:26:51 am	10:37:27 am	03m33s	39°
Ely	08:18:12 am	09:24:15 am	09:27:29 am	10:39:14 am	03m12s	40°
Fallon	08:15:33 am	09:20:02 am	09:22:50 am	10:33:11 am	02m44s	36°
Reno	08:14:55 am	09:18:57 am	09:21:45 am	10:31:40 am	02m45s	35°
Sparks	08:14:58 am	09:18:59 am	09:21:52 am	10:31:45 am	02m51s	35°
Winnemucca	08:16:58 am	09:21:14 am	09:24:40 am	10:34:31 am	03m28s	37°
Oklahoma						
Bartlesville	10:38:45 am	11:52:53 am	11:56:56 am	01:13:49 pm	04m08s	60°
Broken Arrow	10:39:11 am	11:52:58 am	11:58:38 am	01:15:03 pm	05m37s	60°
Del City	10:37:02 am	11:51:52 am	11:54:59 am	01:12:54 pm	03m09s	59°
Edmond	10:36:54 am	11:51:02 am	11:55:16 am	01:12:33 pm	04m13s	59°
Enid	10:36:09 am	11:49:11 am	11:54:41 am	01:10:56 pm	05m34s	58°
Midwest City	10:37:04 am	11:51:48 am	11:55:07 am	01:12:57 pm	03m20s	59°
Moore	10:36:58 am	11:52:22 am	11:54:32 am	01:12:53 pm	02m12s	59°
Muskogee	10:39:54 am	11:53:57 am	11:59:39 am	01:16:13 pm	05m43s	61°
Oklahoma City	10:36:54 am	11:51:32 am	11:54:57 am	01:12:42 pm	03m26s	59°
Ponca City	10:37:13 am	11:50:24 am	11:55:48 am	01:11:58 pm	05m21s	59°
Shawnee	10:37:47 am	11:52:31 am	11:56:13 am	01:13:58 pm	03m42s	60°
Stillwater	10:37:21 am	11:50:46 am	11:56:18 am	01:12:44 pm	05m36s	59°
Tulsa	10:38:59 am	11:52:42 am	11:58:18 am	01:14:42 pm	05m33s	60°
Utah						
Moab	09:22:37 am	10:31:06 am	10:34:33 am	11:48:21 am	03m25s	46°
Murray	09:21:12 am	10:28:30 am	10:30:55 am	11:43:28 am	02m26s	43°
Orem	09:21:12 am	10:27:55 am	10:31:59 am	11:43:55 am	04m05s	43°
Provo	09:21:11 am	10:27:50 am	10:32:10 am	11:44:01 am	04m18s	43°
Salt Lake City	09:21:13 am	10:28:55 am	10:30:30 am	11:43:24 am	01m32s	43°
Sandy City	09:21:25 am	10:28:45 am	10:31:22 am	11:43:56 am	02m40s	43°
South Jordan	09:20:59 am	10:27:45 am	10:31:13 am	11:43:17 am	03m32s	43°
Taylorsville	09:21:09 am	10:28:25 am	10:30:53 am	11:43:23 am	02m28s	43°
West Jordan	09:21:05 am	10:28:08 am	10:31:01 am	11:43:20 am	02m58s	43°
West Valley	09:21:07 am	10:28:24 am	10:30:47 am	11:43:18 am	02m23s	43°

Table 6 – Total Eclipse Circumstances for West Indies and South America

Location	Partial Eclipse Begins	Total Eclipse Begins	Total Eclipse Ends	Partial Eclipse Ends	Duration of Totality	Sun Altitude
Bahamas						
Freeport	12:17:14 pm	01:37:13 pm	01:43:04 pm	02:59:19 pm	05m47s	77°
Nassau	12:22:03 pm	01:42:13 pm	01:48:14 pm	03:04:00 pm	06m01s	77°
Matthew Town	12:36:13 pm	01:56:44 pm	02:02:06 pm	03:16:46 pm	05m25s	74°
Brazil						
Anil	03:04:54 pm	04:08:58 pm	04:12:56 pm	05:10:24 pm	03m57s	26°
Belem	02:58:50 pm	04:06:33 pm	04:08:05 pm	05:08:40 pm	01m30s	31°
Cabo	03:18:44 pm	04:18:44 pm	04:20:15 pm	05:14:37 pm	01m35s	14°
Caico	03:15:34 pm	04:15:41 pm	04:19:18 pm	05:13:38 pm	03m39s	17°
Camarajibe	03:18:31 pm	04:17:51 pm	04:20:39 pm	05:14:27 pm	02m46s	14°
Campina Grande	03:17:10 pm	04:16:41 pm	04:20:07 pm	05:14:03 pm	03m29s	15°
Castanhal	02:59:15 pm	04:05:39 pm	04:09:14 pm	05:08:39 pm	03m35s	31°
Joao Pessoa	03:17:44 pm	04:16:49 pm	04:20:16 pm	05:13:50 pm	03m26s	14°
Recife	03:18:36 pm	04:17:54 pm	04:20:40 pm	05:14:27 pm	02m45s	14°
Sao Luis	03:04:51 pm	04:08:56 pm	04:12:54 pm	05:10:23 pm	03m57s	26°
Vitoria de Santo Antao	03:18:25 pm	04:18:20 pm	04:20:13 pm	05:14:33 pm	01m56s	14°
Dominican Republic						
Benemerita de San Cri.	12:48:00 pm	02:07:23 pm	02:13:10 pm	03:25:41 pm	05m49s	69°
Concepcion de la Vega	12:45:42 pm	02:05:15 pm	02:10:50 pm	03:23:44 pm	05m33s	70°
La Romana	12:50:18 pm	02:10:06 pm	02:13:59 pm	03:26:52 pm	03m54s	68°
San Francisco de Mac.	12:46:06 pm	02:05:48 pm	02:10:44 pm	03:23:51 pm	04m58s	70°
San Juan de la Maguana	12:45:00 pm	02:04:53 pm	02:10:23 pm	03:23:42 pm	05m32s	71°
San Pedro de Macoris	12:49:33 pm	02:09:02 pm	02:13:46 pm	03:26:27 pm	04m47s	68°
Santiago de los Cabal.	12:44:55 pm	02:04:32 pm	02:10:02 pm	03:23:05 pm	05m29s	70°
Santo Domingo	12:48:19 pm	02:07:39 pm	02:13:20 pm	03:25:48 pm	05m40s	69°
French Guiana						
Saul	02:44:58 pm	03:54:50 pm	03:59:10 pm	05:01:30 pm	04m18s	40°
Guyana						
Georgetown	02:31:54 pm	03:44:51 pm	03:49:46 pm	04:54:51 pm	04m56s	48°
Haiti						
Cap-Haitien	01:41:20 pm	03:01:25 pm	03:07:00 pm	04:20:48 pm	05m37s	72°
Gonaives	01:40:54 pm	03:02:14 pm	03:05:51 pm	04:20:47 pm	03m41s	73°
Suriname						
Paramaribo	02:38:21 pm	03:50:38 pm	03:53:12 pm	04:57:43 pm	02m33s	43°
Trinidad & Tobago						
Port of Spain	01:19:06 pm	02:34:28 pm	02:39:23 pm	03:46:52 pm	04m53s	54°
San Fernando	01:19:52 pm	02:35:05 pm	02:40:15 pm	03:47:29 pm	05m11s	54°
Turks & Caicos Is.						
Providenciales Island	11:37:39 am	12:58:43 pm	01:01:44 pm	02:16:49 pm	02m59s	73°
West Caicos Island	11:37:29 am	12:57:58 pm	01:02:17 pm	02:16:52 pm	04m22s	73°
Venezuela						
Carupano	01:15:59 pm	02:32:35 pm	02:36:52 pm	03:45:34 pm	04m15s	56°
Porlamar	01:14:19 pm	02:31:52 pm	02:35:04 pm	03:44:36 pm	03m17s	57°
Tucupita	01:21:09 pm	02:38:07 pm	02:40:02 pm	03:48:57 pm	01m55s	54°

Table 7 - Partial Eclipse Circumstances for the USA

State	City	Partial Eclipse Begins	Maximum Eclipse	Partial Eclipse Ends	Eclipse Magnitude	Sun Altitude
Alabama	Birmingham	10:54:50 am	12:14:34 pm	01:34:05 pm	0.984	69°
	Mobile	10:54:51 am	12:15:55 pm	01:36:50 pm	0.993	71°
Alaska	Anchorage	06:41:55 am	07:22:57 am	08:05:25 am	0.342	15°
Arizona	Phoenix	08:20:26 am	09:29:53 am	10:45:29 am	0.825	45°
	Tucson	08:21:58 am	09:31:53 am	10:47:53 am	0.789	46°
California	Los Angeles	08:15:04 am	09:21:14 am	10:33:52 am	0.826	38°
	Sacramento	08:13:33 am	09:18:23 am	10:29:12 am	0.974	34°
	San Diego	08:16:12 am	09:22:39 am	10:35:37 am	0.782	39°
	San Francisco	08:12:46 am	09:17:12 am	10:27:45 am	0.946	33°
	San Jose	08:12:56 am	09:17:37 am	10:28:28 am	0.932	34°
Colorado	Denver	09:27:30 am	10:39:10 am	11:55:13 am	0.998	50°
Connecticut	Hartford	12:19:45 pm	01:26:38 pm	02:31:38 pm	0.548	62°
District of Columbia	Washington	12:11:12 pm	01:24:02 pm	02:35:06 pm	0.684	66°
Florida	Jacksonville	12:07:00 pm	01:28:29 pm	02:47:45 pm	0.983	74°
	Sarasota	12:08:43 pm	01:31:37 pm	02:52:01 pm	0.998	77°
Georgia	Atlanta	11:59:02 am	01:18:36 pm	02:37:17 pm	0.939	70°
Illinois	Chicago	10:51:40 am	12:04:22 pm	01:17:40 pm	0.760	61°
Indiana	Indianapolis	11:53:57 am	01:08:38 pm	02:23:28 pm	0.800	63°
Iowa	Des Moines	10:42:23 am	11:55:28 am	01:10:28 pm	0.839	58°
Kansas	Wichita	10:36:44 am	11:52:01 am	01:10:16 pm	0.997	58°
Kentucky	Louisville	11:54:45 am	01:10:49 pm	02:26:51 pm	0.836	65°
Louisiana	Baton Rouge	10:49:44 am	12:10:06 pm	01:31:23 pm	0.937	69°
	New Orleans	10:52:04 am	12:12:56 pm	01:34:15 pm	0.940	71°
Maryland	Baltimore	12:11:52 pm	01:24:07 pm	02:34:35 pm	0.669	65°
Massachusetts	Boston	12:23:11 pm	01:28:06 pm	02:31:05 pm	0.510	61°
	Springfield	12:19:57 pm	01:26:21 pm	02:30:53 pm	0.539	62°
Michigan	Detroit	11:59:36 am	01:10:52 pm	02:21:50 pm	0.687	62°
Minnesota	Minneapolis	10:44:21 am	11:54:04 am	01:05:28 pm	0.734	55°
Missouri	Kansas City	10:40:36 am	11:55:35 am	01:12:48 pm	0.924	59°
	St. Louis	10:47:12 am	12:03:03 pm	01:20:01 pm	0.886	62°
Nebraska	Omaha	10:38:56 am	11:51:59 am	01:07:30 pm	0.873	56°
Nevada	Las Vegas	08:17:22 am	09:25:20 am	10:39:23 am	0.905	41°
New Jersey	Newark	12:16:42 pm	01:25:52 pm	02:33:09 pm	0.596	63°
New Mexico	Albuquerque	09:25:33 am	10:37:57 am	11:55:35 am	0.912	50°
New York	Buffalo	12:07:22 pm	01:16:25 pm	02:24:30 pm	0.615	62°
	New York	12:17:03 pm	01:26:08 pm	02:33:18 pm	0.593	63°
North Carolina	Charlotte	12:04:53 pm	01:22:46 pm	02:39:03 pm	0.842	69°
	Winston-Salem	12:05:37 pm	01:22:34 pm	02:37:55 pm	0.809	69°
Ohio	Cincinnati	11:56:51 am	01:11:56 pm	02:26:46 pm	0.795	65°
	Cleveland	12:01:55 pm	01:13:45 pm	02:24:57 pm	0.690	63°
	Columbus	11:59:31 am	01:13:24 pm	02:26:47 pm	0.750	64°
	Dayton	11:57:22 am	01:11:45 pm	02:25:51 pm	0.773	64°
Oregon	Portland	08:17:50 am	09:20:11 am	10:27:17 am	0.866	32°
Pennsylvania	Philadelphia	12:14:43 pm	01:25:26 pm	02:34:15 pm	0.630	64°
	Pittsburgh	12:05:03 pm	01:17:30 pm	02:28:54 pm	0.693	64°
Rhode Island	Providence	12:22:29 pm	01:28:22 pm	02:32:16 pm	0.527	62°
South Carolina	Columbia	12:05:11 pm	01:24:07 pm	02:41:18 pm	0.877	71°
Tennessee	Memphis	10:48:24 am	12:06:49 pm	01:26:09 pm	0.986	66°
	Nashville	10:53:32 am	12:11:25 pm	01:29:21 pm	0.909	67°
Texas	Austin	10:39:39 am	11:57:14 am	01:17:58 pm	0.842	62°
	Dallas	10:39:08 am	11:56:43 am	01:17:13 pm	0.931	62°
	El Paso	09:27:05 am	10:39:38 am	11:57:36 am	0.803	52°
	Houston	10:43:43 am	12:02:33 pm	01:23:43 pm	0.857	66°
	San Antonio	10:39:24 am	11:56:35 am	01:17:07 pm	0.806	62°
Virginia	Richmond	12:10:38 pm	01:25:08 pm	02:37:44 pm	0.726	67°
Washington	Seattle	08:20:13 am	09:21:31 am	10:27:08 am	0.799	32°
West Virginia	Huntington	12:00:42 pm	01:16:02 pm	02:30:35 pm	0.783	66°
Wisconsin	Milwaukee	10:51:31 am	12:03:01 pm	01:15:11 pm	0.731	59°

Table 8 - Partial Eclipse Circumstances for North, Central, and South America

Country	City	Partial Eclipse Begins	Maximum Eclipse	Partial Eclipse Ends	Eclipse Magnitude	Sun Altitude
Barbados	Bridgetown	01:18:18 pm	02:34:58 pm	03:43:55 pm	0.926	53°
Bolivia	La Paz	02:21:06 pm	03:07:08 pm	03:50:14 pm	0.223	42°
Brazil	Rio de Janeiro	03:34:19 pm	04:25:38 pm	05:13:12 pm	0.459	15°
	Sao Paulo	03:35:29 pm	04:24:55 pm	05:10:54 pm	0.395	17°
Canada	Calgary, AB	09:29:16 am	10:30:06 am	11:34:13 am	0.676	37°
	Edmonton, AB	09:33:27 am	10:31:47 am	11:32:52 am	0.596	37°
	Halifax, NS	01:40:01 pm	02:34:42 pm	03:27:35 pm	0.347	56°
	Hamilton, ON	12:05:37 pm	01:14:38 pm	02:22:52 pm	0.619	61°
	Kitchener, ON	12:04:29 pm	01:13:30 pm	02:21:52 pm	0.623	61°
	Laval, QC	12:18:06 pm	01:20:31 pm	02:21:40 pm	0.477	59°
	London, ON	12:03:00 pm	01:12:54 pm	02:22:13 pm	0.645	61°
	Mississauga, ON	12:06:08 pm	01:14:37 pm	02:22:19 pm	0.607	61°
	Montreal, QC	12:18:27 pm	01:20:50 pm	02:21:56 pm	0.476	59°
	North York, ON	12:06:33 pm	01:14:42 pm	02:22:05 pm	0.600	61°
	Oshawa, ON	12:07:39 pm	01:15:22 pm	02:22:14 pm	0.589	61°
	Ottawa, ON	12:14:10 pm	01:18:06 pm	02:20:56 pm	0.508	59°
	Quebec, QC	12:23:41 pm	01:22:22 pm	02:19:45 pm	0.414	57°
	St. Catharines, ON	12:06:44 pm	01:15:35 pm	02:23:33 pm	0.613	61°
	Saskatoon, SK	10:36:29 am	11:37:24 am	12:40:51 pm	0.615	42°
	Scarborough, ON	12:06:49 pm	01:14:57 pm	02:22:16 pm	0.599	61°
	Toronto, ON	12:06:35 pm	01:14:53 pm	02:22:22 pm	0.602	61°
	Vancouver, BC	08:21:58 am	09:21:54 am	10:25:50 am	0.746	31°
	Victoria, BC	08:20:46 am	09:21:12 am	10:25:49 am	0.773	31°
	Windsor, ON	11:59:39 am	01:10:57 pm	02:21:56 pm	0.688	62°
	Winnipeg, MB	10:43:27 am	11:47:31 am	12:53:25 pm	0.624	49°
Colombia	Bogota	12:13:21 pm	01:30:03 pm	02:39:50 pm	0.622	66°
Costa Rica	San Jose	10:44:36 am	12:00:59 pm	01:12:54 pm	0.516	83°
Cuba	Havana	12:15:23 pm	01:39:11 pm	02:59:36 pm	0.892	81°
Ecuador	Quito	12:24:13 pm	01:31:16 pm	02:33:07 pm	0.399	67°
El Salvador	San Salvador	10:25:54 am	11:41:52 am	12:55:30 pm	0.501	85°
Guatemala	Guatemala	10:21:13 am	11:36:54 am	12:50:54 pm	0.500	83°
Haiti	Port-au-Prince	01:43:18 pm	03:06:23 pm	04:22:53 pm	0.994	72°
Honduras	Tegucigalpa	10:27:37 am	11:45:45 am	01:00:49 pm	0.554	88°
Jamaica	Kingston	11:35:43 am	12:59:34 pm	02:17:36 pm	0.882	78°
Mexico	Mexico City	10:55:25 am	12:07:11 pm	01:21:22 pm	0.487	67°
	Guadalajara	10:46:30 am	11:55:16 am	01:07:58 pm	0.473	60°
	Monterrey	10:41:25 am	11:56:37 am	01:15:29 pm	0.666	62°
Nicaragua	Managua	10:34:45 am	11:51:43 am	01:05:05 pm	0.523	87°
Panama	Panama	11:53:39 am	01:12:09 pm	02:24:47 pm	0.598	77°
Paraguay	Asuncion	03:43:22 pm	04:20:35 pm	04:55:56 pm	0.176	26°
Peru	Lima	01:10:53 pm	01:51:26 pm	02:29:55 pm	0.142	54°
Puerto Rico	San Juan	12:56:05 pm	02:16:18 pm	03:29:23 pm	0.946	64°
Venezuela	Caracas	01:09:46 pm	02:30:03 pm	03:42:27 pm	0.929	61°

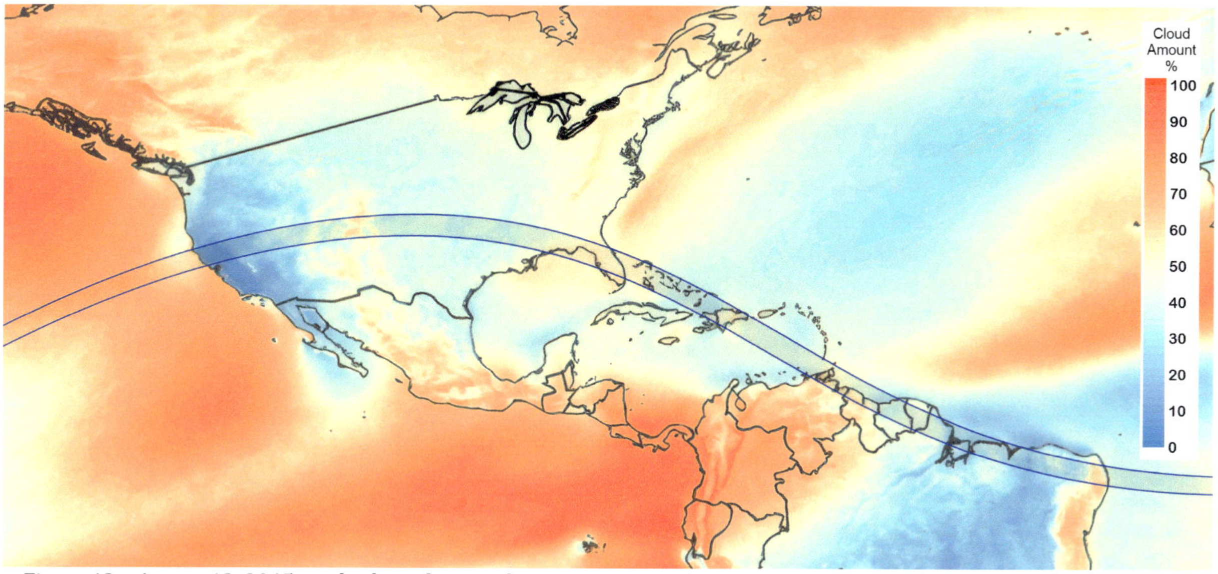

Figure 12 – August 12, 2045 total solar eclipse with average August daytime cloud cover derived from 20 years of satellite data, 2000-2019. Data: CM-SAF. (based on map by Jay Anderson)

5 – Average Cloud Cover for the 2045 Total Solar Eclipse

The average cloud cover along the 2045 total eclipse track is lowest in the western USA. It increases in the eastern USA peaking in Florida. Some locations in the West Indies and South America also offer fair prospects.

Table 9 – Average Cloud Cover and High Temperature in August

State/Country	City	Duration Totality	Ave. High Temp.[1]	Ave. Cloud Cover[2]
California	Redding	04m25s	36°C (96°F)	9%
California	Susanville	04m33s	30°C (84°F)	10%
Nevada	Reno	02m45s	31°C (88°F)	9%
Nevada	Winnemucca	03m28s	32°C (89°F)	12%
Utah	Salt Lake City	01m32s	32°C (89°F)	20%
Utah	Moab	03m25s	32°C (90°F)	29%
Colorado	Grand Junction	05m04s	32°C (89°F)	30%
Colorado	Colorado Springs	05m04s	27°C (80°F)	40%
Oklahoma	Enid	05m34s	33°C (92°F)	45%
Oklahoma	Tulsa	05m33s	33°C (92°F)	44%
Arkansas	Fort Smith	05m44s	33°C (92°F)	39%
Arkansas	Little Rock	05m36s	33°C (91°F)	58%
Mississippi	Jackson	03m10s	32°C (90°F)	61%
Mississippi	Meridian	05m51s	32°C (89°F)	59%
Alabama	Dothan	05m40s	31°C (88°F)	71%
Florida	Panama City	05m10s	31°C (87°F)	69%
Florida	Tallahassee	05m47s	32°C (89°F)	65%
Florida	Gainesville	04m36s	31°C (88°F)	68%
Florida	Orlando	05m42s	32°C (89°F)	72%
Florida	Miami	02m42s	31°C (88°F)	74%
Bahamas	Freeport	05m47s	30°C (86°F)	60%
Bahamas	Nassau	06m01s	31°C (88°F)	63%
Dominican Republic	Santo Domingo	05m40s	31°C (88°F)	80%
Trinidad & Tabago	San Fernando	05m11s	31°C (87°F)	80%
Guyana	Georgetown	04m56s	30°C (86°F)	50%
Suriname	Paramaribo	02m33s	31°C (88°F)	40%
Brazil	Belem	01m30s	32°C (89°F)	49%
Brazil	Sao Luis	03m57s	31°C (88°F)	41%
Brazil	Caico	03m39s	35°C (95°F)	35%
Brazil	Joao Pessoa	03m26s	28°C (82°F)	70%

[1]*Average High Temperature in August [source: www.weatherspark.com]*

[2]*Average Cloud Cover in August (since 2000) [source: www.timeanddate.com]*

6 – Saros 136

The total eclipse of 2045 is part of a series of eclipses with the longest totalities of the 20th and 21st centuries. These remarkable eclipses all have durations exceeding six minutes and belong to a group called Saros 136.

The Saros is a period in which eclipses repeat with very similar characteristics. It has a duration of approximately 6,585.32 days (~18 years 11 days 8 hours). When two eclipses are separated by a period of one Saros, they share a very similar geometry. The eclipses occur at the same node with the Moon at nearly the same distance from Earth and at the same time of year. Thus, the Saros is useful for organizing eclipses into families or series. A Saros series doesn't last forever because the characteristics of the eclipse geometry slowly change with each cycle. A series typically lasts 12 or 13 centuries and contains 70 or more eclipses.

The total eclipse of 2045 is the 39th member of Saros series 136. The Saros series were first numbered by van den Bergh in 1955. All eclipses in an even numbered Saros series occur at the Moon's descending node and the Moon moves northward with each succeeding member in the family (i.e., gamma[6] increases). Saros 136 is a vigorous series in the prime of its life. The series began with a small partial eclipse visible off the coast of Antarctica on Jun 14, 1360. After seven more partial eclipses, each of increasing magnitude, the first umbral eclipse occurred on Sep 8, 1504. This event was a central annular eclipse of short duration that was visible from Antarctica and the South Pacific Ocean.

The next five members of Saros 136 were all annular eclipses; the duration of annularity steadily decreased with each eclipse as the Moon passed progressively closer to Earth. The character of the series changed to hybrid (also called annular-total) with the eclipse of Nov 22, 1612. The classification of such an eclipse changes from annular to total or vice versa along different portions of the ground track. This dual nature arises due to the curvature of Earth's surface, which brings the middle part of the path into the umbra (total eclipse) while other, more distant segments remain within the antumbral shadow (annular eclipse). Hybrid eclipses are rather rare and account for only 4.8% of the 11,898 solar eclipses occurring during the 5000-year period from 2000 BCE to 3000 CE.

Detailed calculations reveal that the 1612 hybrid eclipse did not actually have a total phase. At greatest eclipse the Sun and Moon subtended virtually the same angular diameter, so the Moon's irregular profile produced a beaded annularity not unlike the solar eclipse of Oct 3, 1986. Beginning with the eclipse of Dec 4, 1630, the following five events were true hybrids with increasing durations of totality. The last of the hybrids in this group occurred on Jan 17, 1703. It is interesting to note that the entire path was total except for the final ~1500 km (930 mi) which was annular. In comparison, most hybrid eclipses are annular at both ends of their paths.

The first purely total eclipse of the series occurred on Jan 27, 1721 and had a maximum duration of 1m 07s. The central track extended from the South Pacific to the South Atlantic, crossing southernmost South America in the process. Succeeding eclipses in the series had increasingly longer durations as the paths shifted north. The total eclipse of Mar 24, 1811 had a maximum duration of 3m 27s, while the eclipse of May 18, 1901 was nearly twice as long at 6m 29s.

Throughout the 20th century, Saros 136 continued to produce exceptionally long total eclipses with tracks passing across the equator. The eclipse of May 29, 1919 is significant because it was the first eclipse used to measure the gravitational deflection of starlight by the Sun as predicted by Einstein's general theory of relativity.

On Jun 8, 1937, the central duration of totality exceeded 7 minutes for the first time for any total eclipse since the year 1098. Unfortunately, most of the path was over the Pacific Ocean. The following eclipse (Jun 20, 1955) was 4 seconds longer with a maximum duration of 7m 08s. Its track crossed Sri Lanka, Southeast Asia, and the Philippines. This event marked the maximum duration of all total eclipses belonging to Saros 136.

Subsequent members in the series are shorter as the Moon recedes farther from perigee. Nevertheless, Saros 136 will continue to produce long eclipses for some time to come. The first successor to the 1955 eclipse occurred on Jun 30, 1973 and lasted 7m 04s from Africa. It was followed by the total eclipse of Jul 11, 1991 with a duration of 6m 53s and a path through Hawaii, Mexico, Central and South America. The next eclipse of the

---

[6] *Gamma is the distance of the Moon's shadow axis from Earth's center in units of equatorial Earth radii. It is defined at the instant of greatest eclipse when its absolute value is at a minimum.*

series occurred on Jul 22, 2009 with a duration of 6m 39s from China and the Pacific Ocean. The upcoming eclipse of Aug 2, 2027 has a duration of 6m 39s and is visible from northern Africa and the Arabian Peninsula.

In comparison, the Aug 12, 2045 total eclipse lasts a maximum of 6m 06s. It is noteworthy because its track passes over the United States from northern California to central Florida where the event will be witnessed by millions of people.

The paths of previous eclipses have steadily shifted north, but this trend reverses starting with the total eclipse of Aug 24, 2063. The retrograde effect is due to the inclination of Earth's axis as the date of each eclipse shifts ~11 days forward through the fall–winter season. The temporary downward shift is the result of the tipping of Earth's Northern Hemisphere away from the Sun, which is greater than the northern shift of the Moon with respect to the node. Nearly two centuries later, the path resumes its northern shift with the eclipse of Dec 12, 2243. By this time, the maximum duration of totality will drop to 3m 30s.

The series continues to produce total eclipses for two more centuries as the duration of totality gradually dwindles. The last total eclipse occurs on May 13, 2496 and has a maximum duration of 1m 02s. The final seven eclipses in Saros 136 are all partial events visible from the Northern Hemisphere. The series terminates with a partial eclipse visible primarily from eastern Siberia on Jul 30, 2622.

In summary, Saros series 136 includes 71 eclipses. It begins with 8 partial eclipses, followed by 6 annular, 6 hybrid, and 44 total eclipses. It ends with a string of 7 partial eclipses. From start to finish, the series spans a period of 1262 years. The EclipseWise.com catalog for Saros 136 has more details and maps for this series:

https://www.eclipsewise.com/solar/SEsaros/SEsaros136.html

Recent Total Solar Eclipses of Saros 136

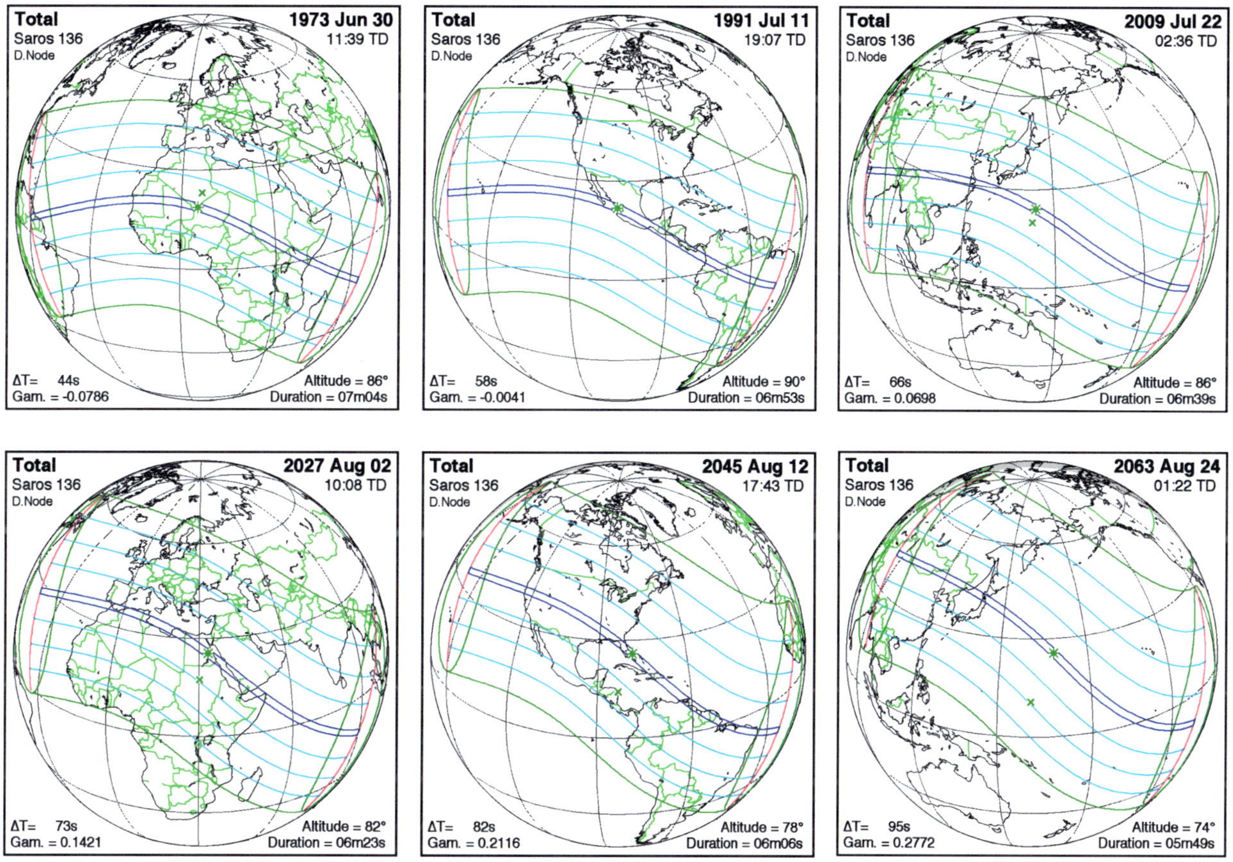

Figure 13 – Six total solar eclipses of Saros 136 for the years 1973 (7m 04s), 1991 (6m 53s), 2009 (6m 39s), 2045 (6m 23s), 2045 (6m 06s), and 2063 (5m 49s). Note how the eclipse path returns to approximately the same geographic region after 54 years, 33 days (3 cycles). This period of three Saros's is called an Exeligmos.

7 – Detailed Maps of the Path of Totality

A series of 37 full-page maps (pages 24-60) cover the entire 2045 path of totality through the USA, the West Indies, and South America. The *Table of Contents* (page iii) can be used to quickly navigate to the map of interest since it lists the region featured on each map.

The map scale is approximately 1:1,400,000, which corresponds to 1 inch ≈ 22 miles (1 cm ≈ 14 km).[7] This large scale shows both major and minor roads, towns and cities, rivers, lakes, parks, national forests, wilderness areas and mountain ranges. A 50-mile (50-km) reference scale appears at the bottom of each map.

The path of totality on each map is depicted as a lightly shaded region with the northern and southern limits clearly labeled. The total eclipse can be seen only inside this path (a partial eclipse is visible outside the path). The closer one gets to the central line, the longer the total eclipse lasts. Gray lines inside the path mark the duration of the total eclipse in 30-second steps. This makes it easy to estimate the duration of totality from any location in the eclipse path.

The local time of mid-eclipse is marked by a series of white lines crossing the eclipse path every 5 minutes. Abbreviations for local times are: PDT, MDT, CDT, EDT, AST, VST, GYT, SRT, and BRT (see Table 1). Eclipse circumstances on the central line are labeled with the local time of mid-eclipse, duration of totality (minutes and seconds) and altitude of the Sun.

All maps were produced using Google Maps as the underlying map with overlying eclipse graphics generated using JavaScript code. A web page is available to the user for examining any part of the 2045 eclipse path at a range of zoom magnifications. An added benefit of the web page is that it automatically calculates the local circumstances for any point chosen by the user, with just the click of a mouse. For more information and to access the interactive 2045 eclipse path plotted on Google Maps, visit:

https://www.eclipsewise.com/solar/SEgmapx/2001-2100/SE2045Aug12Tgmapx.html#map

8 – EclipseWise.com Web Site

EclipseWise.com has individual web pages, maps and diagrams for every solar and lunar eclipse covering the 5000-year period from 2000 BCE to 3000 CE. This includes 11,898 solar eclipses and 12,064 lunar eclipses. Much of the design, layout and graphics were inspired by the recent publications *Thousand Year Canon of Solar Eclipses 1501 to 2500* and the *Thousand Year Canon of Lunar Eclipses 1501 to 2500.* For more information about these two books, visit: ***https://www.astropixels.com/pubs***

The graphical user interface used by *EclipseWise.com* offers an intuitive way of accessing eclipse predictions. For example, the home page presents a concise preview of all upcoming solar and lunar eclipses over several years. Each small eclipse diagram gives a quick preview of an eclipse and links to a dedicated page for that eclipse.

The main or top pages of EclipseWise.com are:

Home Page (both solar and lunar eclipses): ***https://www. eclipsewise.com/eclipse.html***
Solar Eclipses Page: ***https://www. eclipsewise.com/solar/solar.html***
Lunar Eclipses Page: ***https://www. eclipsewise.com/lunar/lunar.html***

---

[7] Because of the Mercator map projection, the actual scale on a given map can vary by up to 10% of this value.

9 – EclipseWise.com and the 2045 Eclipse

EclipseWise.com has a series of pages and resources devoted to the 2045 eclipse. The main page is located at:

https://www.eclipsewise.com/solar/SEprime/2001-2100/SE2045Aug12Tprime.html

It provides links to detailed eclipse path maps, tables of eclipse path coordinates, Besselian elements, and more. The link to an interactive Google Map with the eclipse path plotted on it allows the user to zoom into a part of the path. Click on any point on the map to display the eclipse circumstances and duration of totality at that location.

Other features include information on eye safety, eclipse photography, the sky during totality, and additional data tables about the eclipse path. This web site will continue to add features as the eclipse approaches.

The *2045 Total Solar Eclipse Circumstances Calculator* is an interactive web page that can quickly calculate the local circumstances for the eclipse from any geographic location. The *Calculator* is located at:

https://www.eclipsewise.com/solar/SEcirc/2001-2100/SE2045Aug12Tcirc.html

10 – Eclipse Predictions

The algorithms and software for the eclipse predictions were developed primarily from the *Explanatory Supplement to the Astronomical Ephemeris* (Her Majesty's Nautical Almanac Office, 1974) with additional algorithms from *Elements of Solar Eclipses: 1951–2200* (Meeus, 1989). The solar and lunar ephemerides were generated from the JPL DE430. All eclipse calculations were made using a value for the Moon's radius of k=0.2722810 for the path of totality. Eclipse contact times for locations and cities (Tables 3 through 6) have been corrected for the effects of the Moon's limb profile. A value for ΔT of 72.2 seconds was used to convert the predictions from Terrestrial Time (TT) to Coordinated Universal Time (UTC).

Photo 3 – Viewing the Total Solar Eclipse of August 1, 2008 from Jinta, China.
© 2008 F. Espenak, www.MrEclipse.com

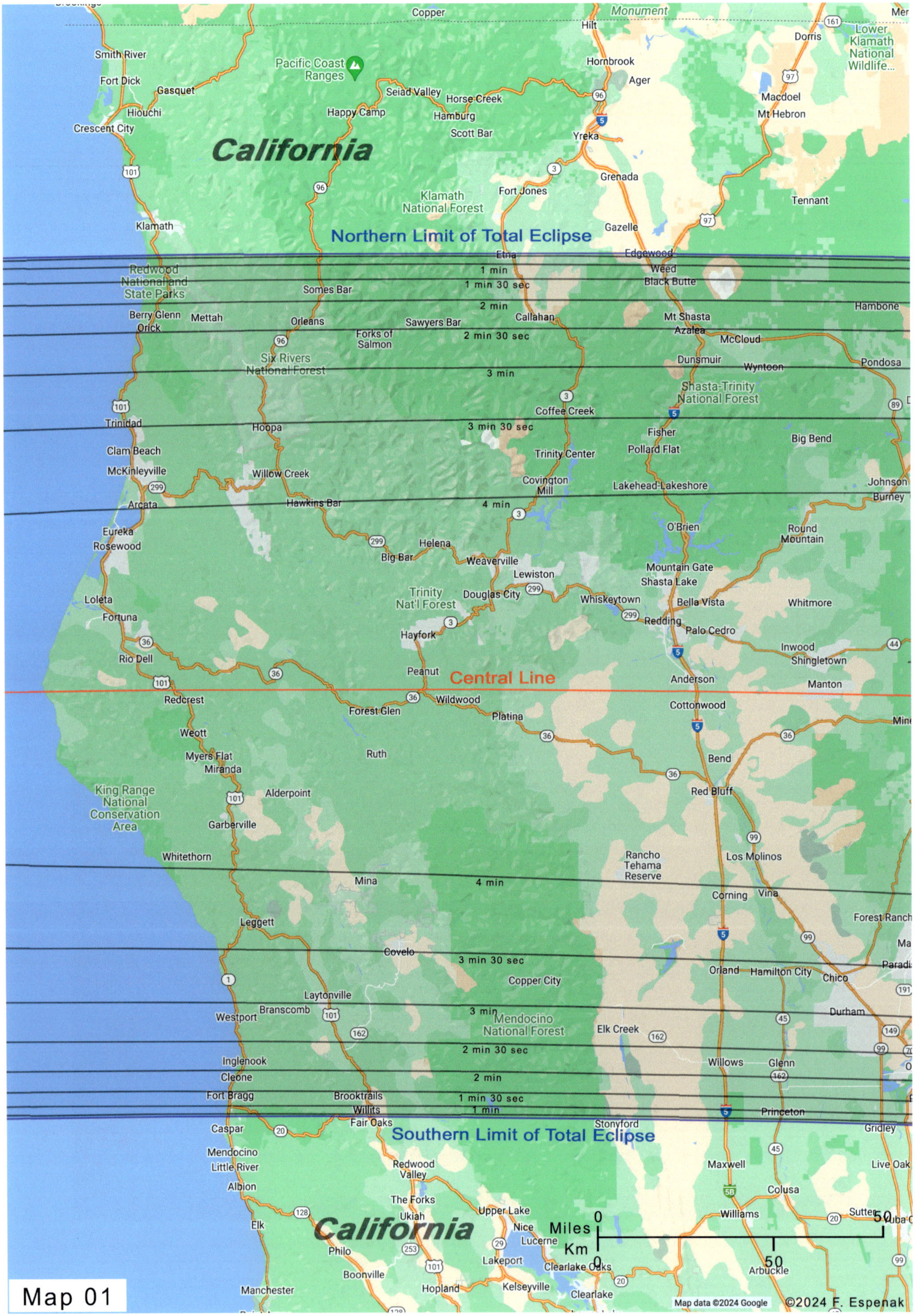
California
Northern Limit of Total Eclipse
1 min
1 min 30 sec
2 min
2 min 30 sec
3 min
3 min 30 sec
4 min
Central Line
4 min
3 min 30 sec
3 min
2 min 30 sec
2 min
1 min 30 sec
1 min
Southern Limit of Total Eclipse
California
Copper
Monument
Hilt
Dorris
Lower Klamath National Wildlife...
Smith River
Fort Dick
Gasquet
Hiouchi
Crescent City
Pacific Coast Ranges
Seiad Valley
Horse Creek
Happy Camp
Hamburg
Scott Bar
Hornbrook
Ager
Macdoel
Mt Hebron
Yreka
Grenada
Klamath
Klamath National Forest
Fort Jones
Gazelle
Tennant
Etna
Edgewood
Weed
Black Butte
Redwood National and State Parks
Somes Bar
Berry Glenn
Orick
Mettah
Orleans
Sawyers Bar
Callahan
Mt Shasta
Azalea
McCloud
Hambone
Forks of Salmon
Six Rivers National Forest
Dunsmuir
Wyntoon
Pondosa
Shasta-Trinity National Forest
Coffee Creek
Trinidad
Hoopa
Fisher
Pollard Flat
Big Bend
Clam Beach
McKinleyville
Willow Creek
Trinity Center
Covington Mill
Lakehead-Lakeshore
Johnson
Burney
Arcata
Hawkins Bar
Eureka
Rosewood
O'Brien
Round Mountain
Helena
Big Bar
Weaverville
Lewiston
Mountain Gate
Shasta Lake
Loleta
Trinity Nat'l Forest
Douglas City
Whiskeytown
Bella Vista
Whitmore
Fortuna
Redding
Palo Cedro
Hayfork
Inwood
Shingletown
Rio Dell
Peanut
Anderson
Manton
Wildwood
Redcrest
Forest Glen
Platina
Cottonwood
Weott
Myers Flat
Miranda
Ruth
Bend
Red Bluff
King Range National Conservation Area
Alderpoint
Garberville
Whitethorn
Rancho Tehama Reserve
Los Molinos
Mina
Corning
Vina
Forest Ranch
Leggett
Covelo
Copper City
Orland
Hamilton City
Chico
Laytonville
Branscomb
Westport
Mendocino National Forest
Elk Creek
Durham
Inglenook
Cleone
Willows
Glenn
Fort Bragg
Brooktrails
Willits
Princeton
Fair Oaks
Caspar
Stonyford
Gridley
Mendocino
Little River
Redwood Valley
Maxwell
Live Oak
Albion
The Forks
Colusa
Ukiah
Upper Lake
Williams
Sutter
Elk
Nice
Lucerne
Philo
Lakeport
Clearlake Oaks
Arbuckle
Boonville
Manchester
Hopland
Kelseyville
Clearlake
Miles 0 50
Km 0 50
Map 01
Map data ©2024 Google
©2024 F. Espenak

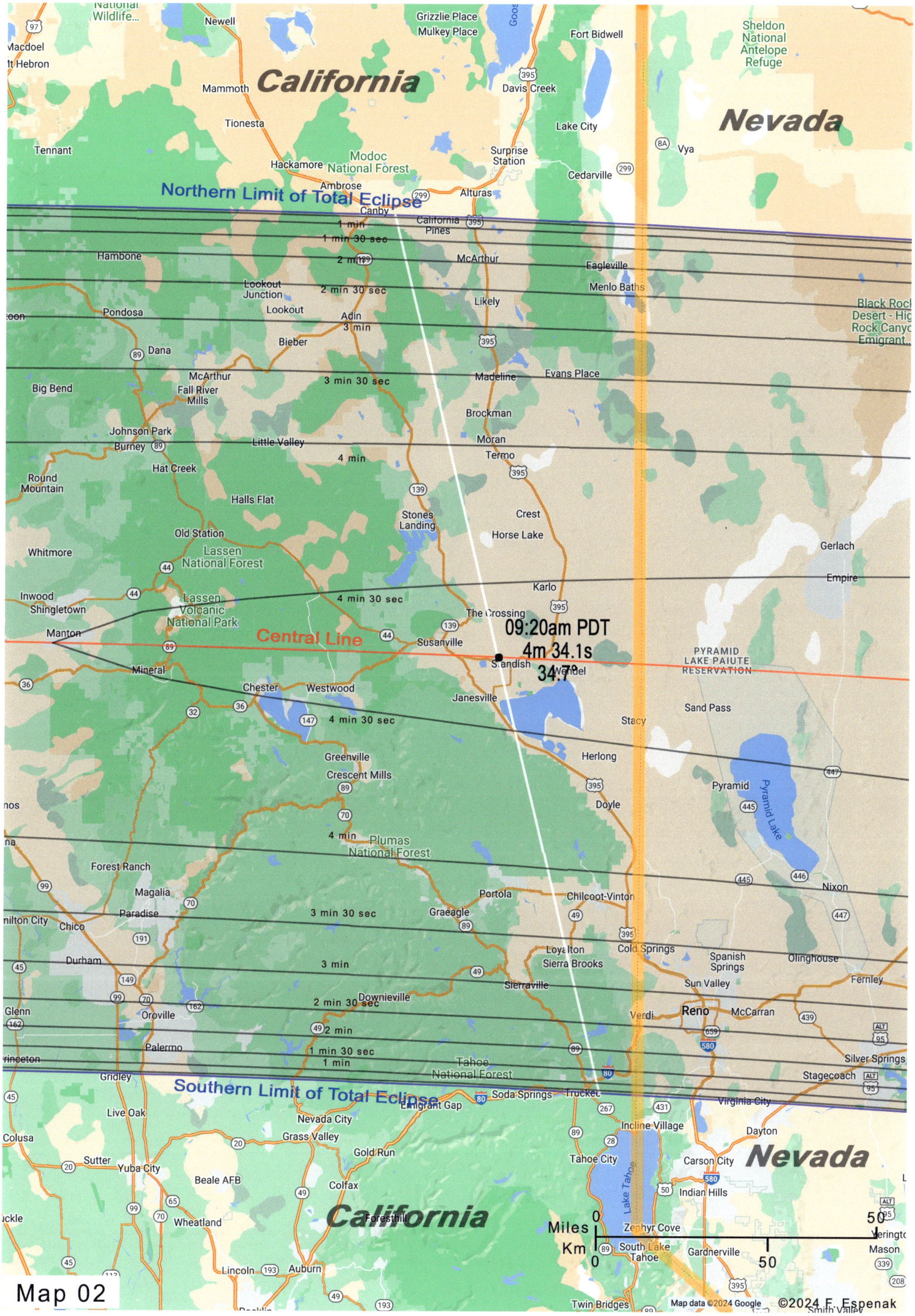
California
Nevada
Northern Limit of Total Eclipse
Southern Limit of Total Eclipse
Central Line
09:20am PDT
4m 34.1s
34.7°
1 min
1 min 30 sec
2 min
2 min 30 sec
3 min
3 min 30 sec
4 min
4 min 30 sec
Modoc National Forest
Lassen National Forest
Lassen Volcanic National Park
Plumas National Forest
Tahoe National Forest
Sheldon National Antelope Refuge
PYRAMID LAKE PAIUTE RESERVATION
Pyramid Lake
Lake Tahoe
Alturas
Canby
Adin
Bieber
Susanville
Standish
Wendel
Janesville
Chester
Westwood
Greenville
Portola
Reno
Sparks
Carson City
Truckee
Tahoe City
South Lake Tahoe
Chico
Oroville
Paradise
Yuba City
Miles
Km
0
50
Map 02
Map data ©2024 Google
©2024 F. Espenak

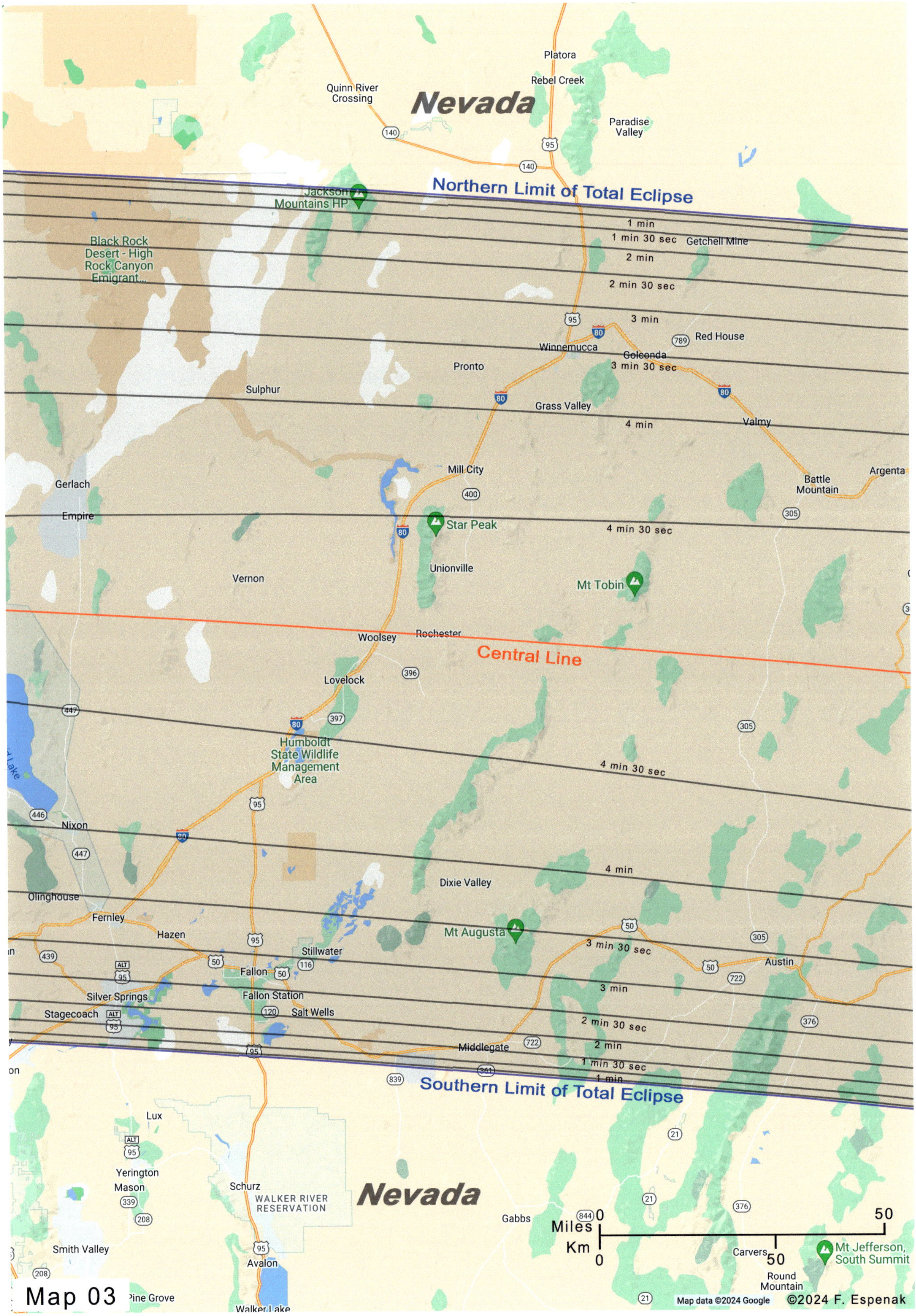
Nevada
Platora
Rebel Creek
Quinn River
Crossing
Paradise
Valley
Northern Limit of Total Eclipse
Jackson
Mountains HP
Black Rock
Desert - High
Rock Canyon
Emigrant...
1 min
1 min 30 sec
Getchell Mine
2 min
2 min 30 sec
3 min
Red House
Winnemucca
Golconda
3 min 30 sec
Pronto
Sulphur
Grass Valley
Valmy
4 min
Mill City
Battle
Mountain
Argenta
Gerlach
Empire
Star Peak
4 min 30 sec
Unionville
Vernon
Mt Tobin
Woolsey
Rochester
Central Line
Lovelock
Humboldt
State Wildlife
Management
Area
4 min 30 sec
Nixon
Olinghouse
Fernley
Hazen
4 min
Dixie Valley
Mt Augusta
3 min 30 sec
Stillwater
Fallon
Fallon Station
Salt Wells
Austin
Silver Springs
Stagecoach
3 min
2 min 30 sec
Middlegate
2 min
1 min 30 sec
1 min
Southern Limit of Total Eclipse
Lux
Yerington
Mason
Schurz
WALKER RIVER
RESERVATION
Nevada
Gabbs
Miles 0 50
Km 0 50
Carvers
Round
Mountain
Mt Jefferson,
South Summit
Smith Valley
Avalon
Pine Grove
Walker Lake
Map 03
Map data ©2024 Google
©2024 F. Espenak

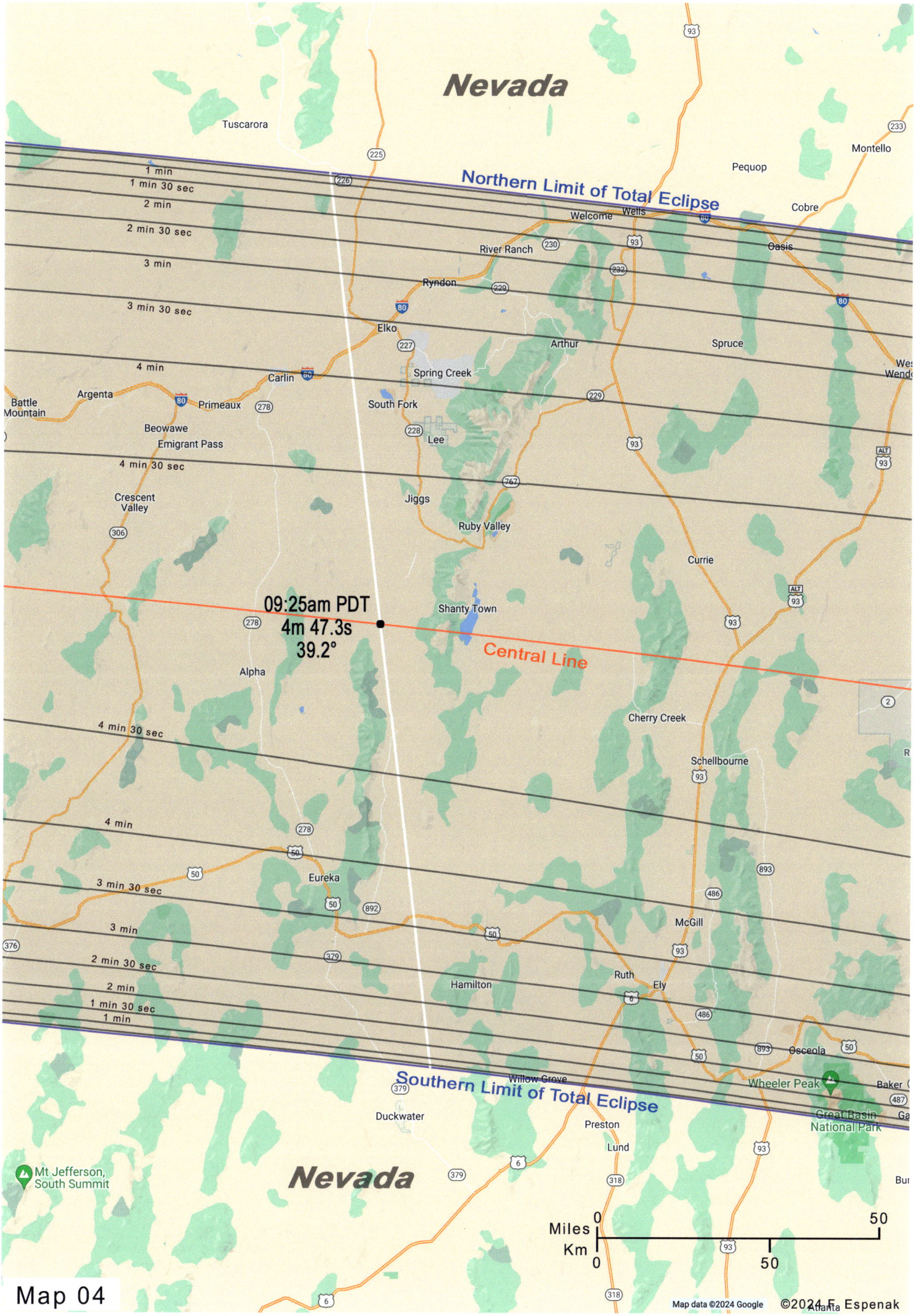
Nevada
Northern Limit of Total Eclipse
Central Line
Southern Limit of Total Eclipse
09:25am PDT
4m 47.3s
39.2°
1 min
1 min 30 sec
2 min
2 min 30 sec
3 min
3 min 30 sec
4 min
4 min 30 sec
Tuscarora
Pequop
Montello
Cobre
Welcome
Wells
Oasis
River Ranch
Ryndon
Elko
Arthur
Spruce
Spring Creek
Carlin
Argenta
Battle Mountain
Primeaux
South Fork
Beowawe
Emigrant Pass
Lee
Jiggs
Ruby Valley
Crescent Valley
Currie
Shanty Town
Alpha
Cherry Creek
Schellbourne
Eureka
McGill
Hamilton
Ruth
Ely
Osceola
Wheeler Peak
Baker
Great Basin National Park
Willow Grove
Duckwater
Preston
Lund
Mt Jefferson, South Summit
Nevada
Miles
Km
0
50
Map 04
Map data ©2024 Google
©2024 F. Espenak

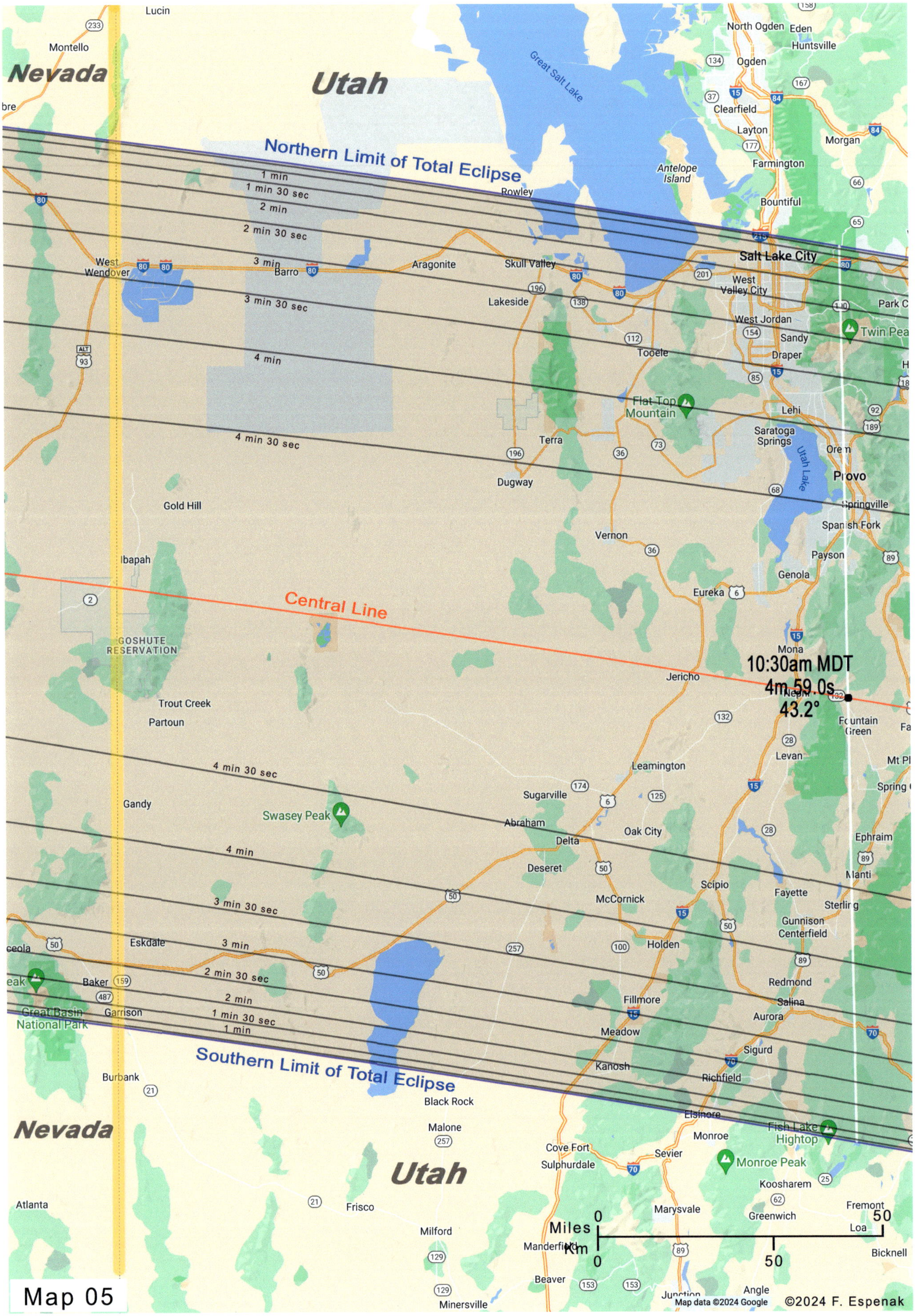
Nevada
Utah
Great Salt Lake
Northern Limit of Total Eclipse
1 min
1 min 30 sec
2 min
2 min 30 sec
3 min
3 min 30 sec
4 min
4 min 30 sec
Central Line
10:30am MDT
4m 59.0s
43.2°
4 min 30 sec
4 min
3 min 30 sec
3 min
2 min 30 sec
2 min
1 min 30 sec
1 min
Southern Limit of Total Eclipse
Lucin
Montello
West Wendover
Barro
Aragonite
Skull Valley
Lakeside
Rowley
Antelope Island
North Ogden
Eden
Huntsville
Ogden
Clearfield
Layton
Morgan
Farmington
Bountiful
Salt Lake City
West Valley City
West Jordan
Sandy
Draper
Park C
Twin Pea
Tooele
Flat Top Mountain
Lehi
Saratoga Springs
Utah Lake
Orem
Provo
Springville
Spanish Fork
Payson
Genola
Terra
Dugway
Vernon
Eureka
Gold Hill
Ibapah
GOSHUTE RESERVATION
Mona
Jericho
Nephi
Fountain Green
Levan
Trout Creek
Partoun
Leamington
Sugarville
Abraham
Oak City
Delta
Deseret
Gandy
Swasey Peak
Mt Pl
Spring
Ephraim
Manti
Scipio
Fayette
McCornick
Sterling
Gunnison
Centerfield
Holden
Eskdale
Baker
Garrison
Great Basin National Park
Fillmore
Redmond
Salina
Aurora
Meadow
Kanosh
Sigurd
Richfield
Elsinore
Monroe
Fish Lake Hightop
Monroe Peak
Burbank
Black Rock
Malone
Cove Fort
Sulphurdale
Sevier
Koosharem
Marysvale
Greenwich
Fremont
Loa
Bicknell
Atlanta
Frisco
Milford
Manderfield
Beaver
Minersville
Junction
Angle
Miles
Km
0
50
Map data ©2024 Google
©2024 F. Espenak
Map 05

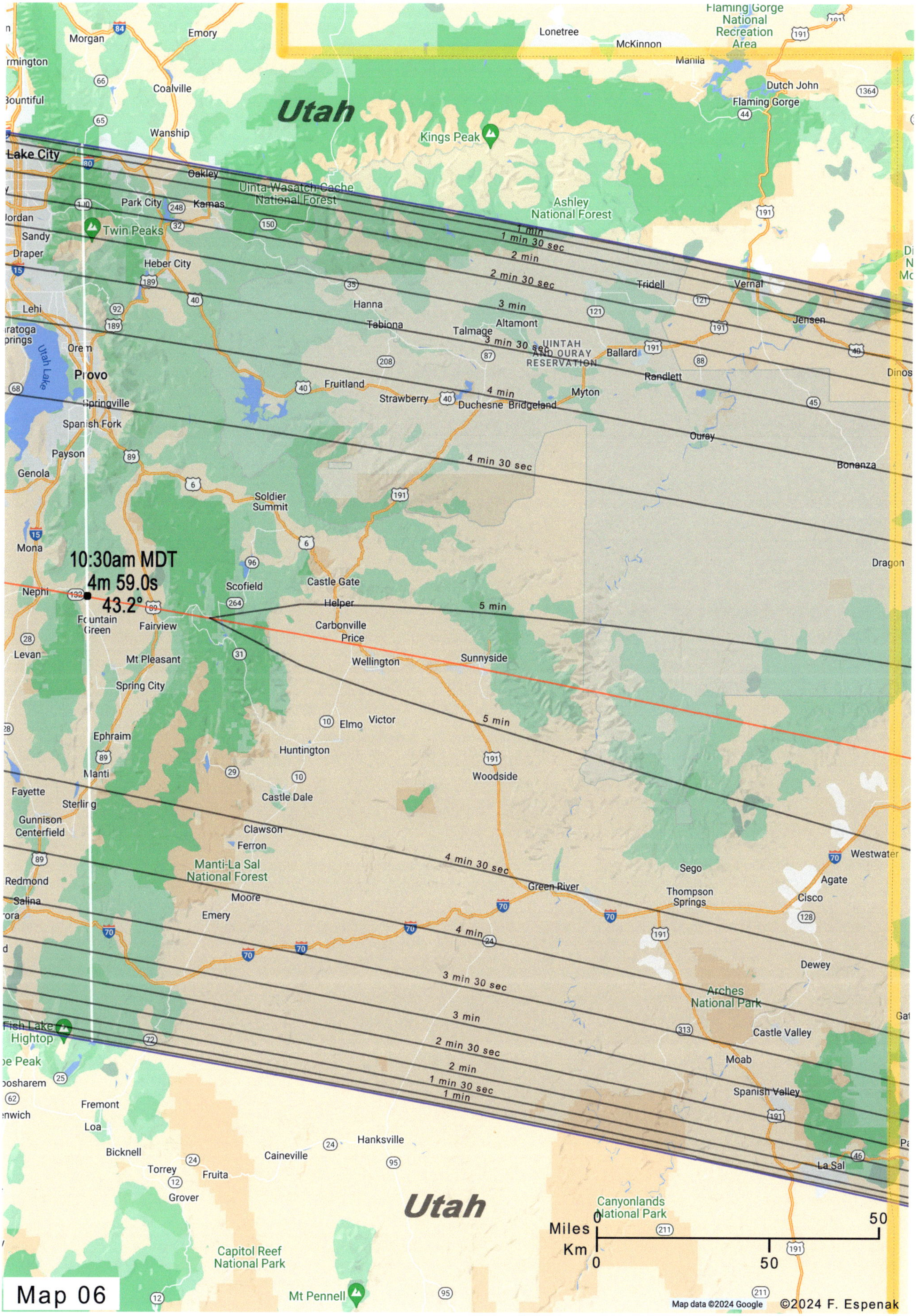
Utah
10:30am MDT
4m 59.0s
43.2°
1 min
1 min 30 sec
2 min
2 min 30 sec
3 min
3 min 30 sec
4 min
4 min 30 sec
5 min
Kings Peak
Uinta-Wasatch-Cache National Forest
Ashley National Forest
Flaming Gorge National Recreation Area
UINTAH AND OURAY RESERVATION
Manti-La Sal National Forest
Arches National Park
Canyonlands National Park
Capitol Reef National Park
Mt Pennell
Provo
Price
Green River
Moab
Vernal
Utah
Miles
Km
Map 06
Map data ©2024 Google
©2024 F. Espenak

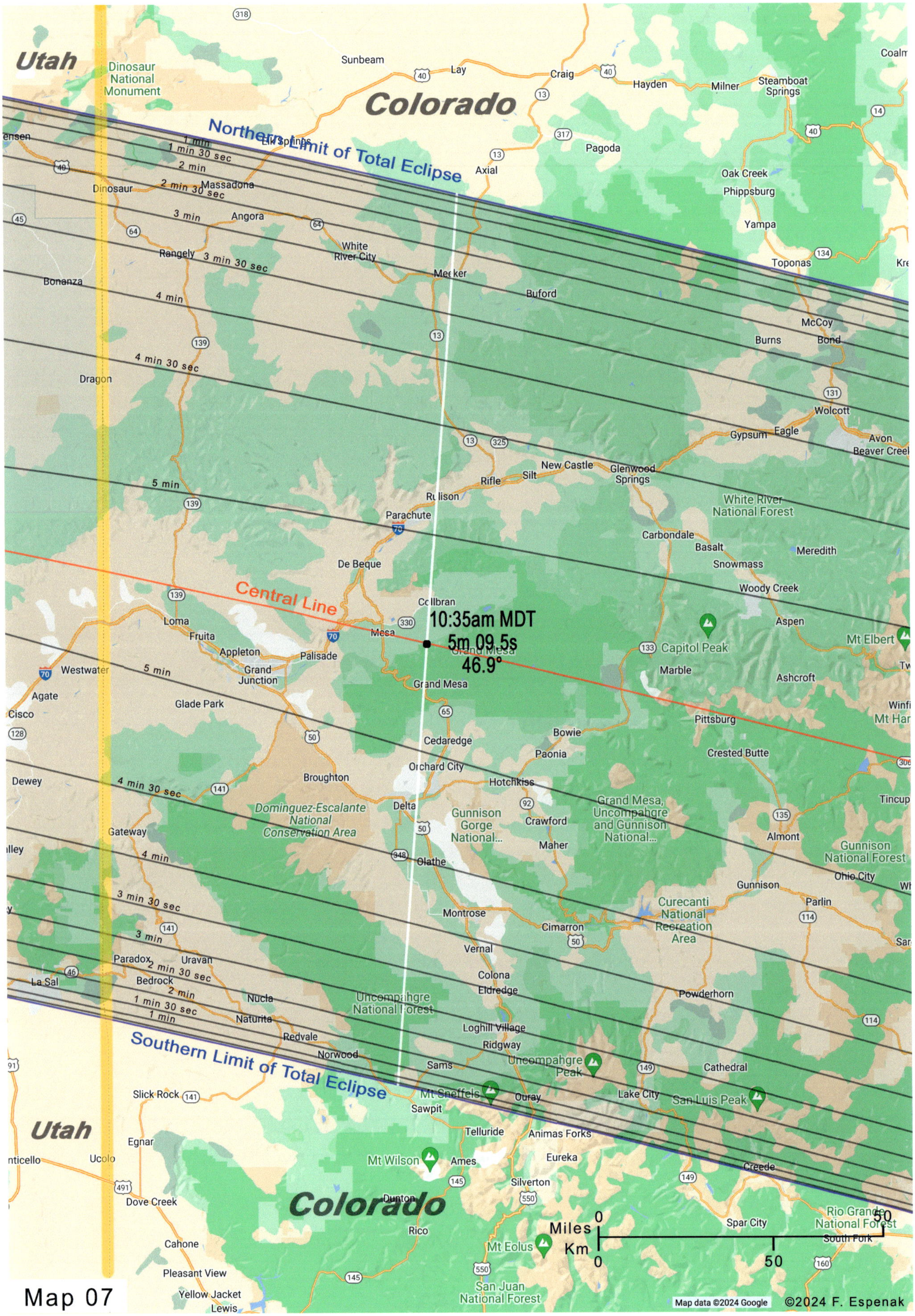
Utah
Colorado
Northern Limit of Total Eclipse
Central Line
Southern Limit of Total Eclipse
10:35am MDT
5m 09.5s
46.9°
1 min
1 min 30 sec
2 min
2 min 30 sec
3 min
3 min 30 sec
4 min
4 min 30 sec
5 min
Dinosaur National Monument
Sunbeam
Lay
Craig
Hayden
Milner
Steamboat Springs
Elk Springs
Pagoda
Axial
Oak Creek
Phippsburg
Yampa
Toponas
Massadona
Dinosaur
Angora
White River City
Rangely
Meeker
Buford
Bonanza
McCoy
Bond
Burns
Dragon
Wolcott
Gypsum
Eagle
Avon
Beaver Creek
New Castle
Glenwood Springs
Silt
Rifle
Rulison
Parachute
White River National Forest
Carbondale
Basalt
Snowmass
Meredith
Woody Creek
De Beque
Aspen
Collbran
Mt Elbert
Loma
Fruita
Mesa
Capitol Peak
Appleton
Palisade
Grand Mesa
Marble
Westwater
Grand Junction
Ashcroft
Agate
Cisco
Glade Park
Pittsburg
Bowie
Cedaredge
Paonia
Crested Butte
Orchard City
Hotchkiss
Dewey
Broughton
Dominguez-Escalante National Conservation Area
Delta
Gunnison Gorge National...
Crawford
Grand Mesa, Uncompahgre and Gunnison National...
Tincup
Almont
Gateway
Maher
Gunnison National Forest
Olathe
Ohio City
Gunnison
Parlin
Montrose
Curecanti National Recreation Area
Cimarron
Paradox
Uravan
Vernal
Colona
Eldredge
La Sal
Bedrock
Nucla
Naturita
Powderhorn
Uncompahgre National Forest
Loghill Village
Redvale
Ridgway
Norwood
Sams
Uncompahgre Peak
Cathedral
Mt Sneffels
Ouray
Lake City
San Luis Peak
Slick Rock
Sawpit
Telluride
Animas Forks
Egnar
Eureka
Creede
Mt Wilson
Ames
Ucolo
Silverton
Dove Creek
Dunton
Rio Grande National Forest
Spar City
Rico
South Fork
Miles
Km
0
50
Mt Eolus
Cahone
Pleasant View
Yellow Jacket
Lewis
San Juan National Forest
Map 07
Map data ©2024 Google
©2024 F. Espenak

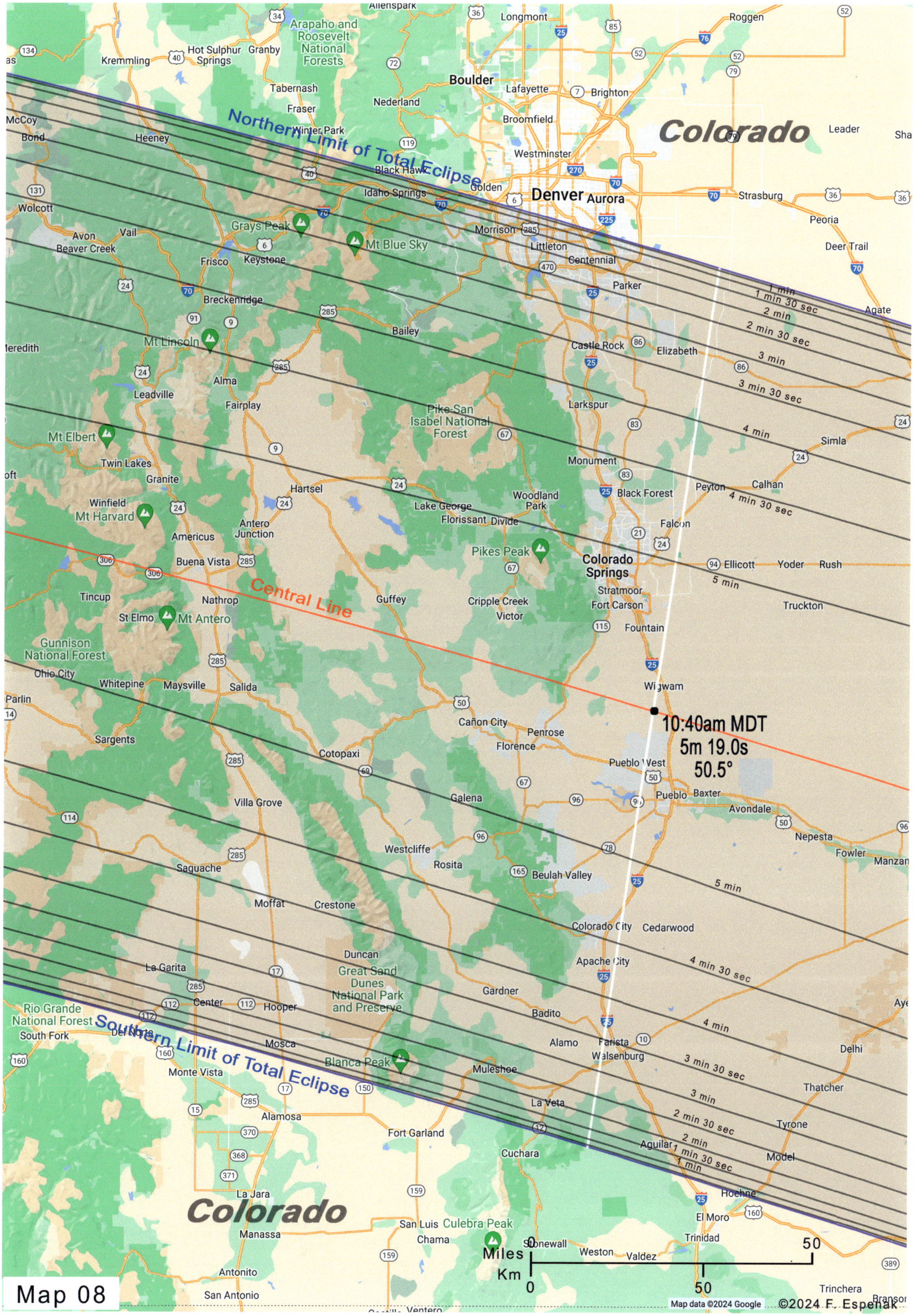
Northern Limit of Total Eclipse
Central Line
Southern Limit of Total Eclipse
10:40am MDT
5m 19.0s
50.5°
1 min
1 min 30 sec
2 min
2 min 30 sec
3 min
3 min 30 sec
4 min
4 min 30 sec
5 min
Colorado
Denver
Boulder
Colorado Springs
Pueblo
Pikes Peak
Mt Elbert
Mt Harvard
Mt Antero
Mt Lincoln
Grays Peak
Mt Blue Sky
Blanca Peak
Culebra Peak
Great Sand Dunes National Park and Preserve
Pike-San Isabel National Forest
Gunnison National Forest
Rio Grande National Forest
Arapaho and Roosevelt National Forests
Miles
Km
0
50
Map 08
Map data ©2024 Google
©2024 F. Espenak

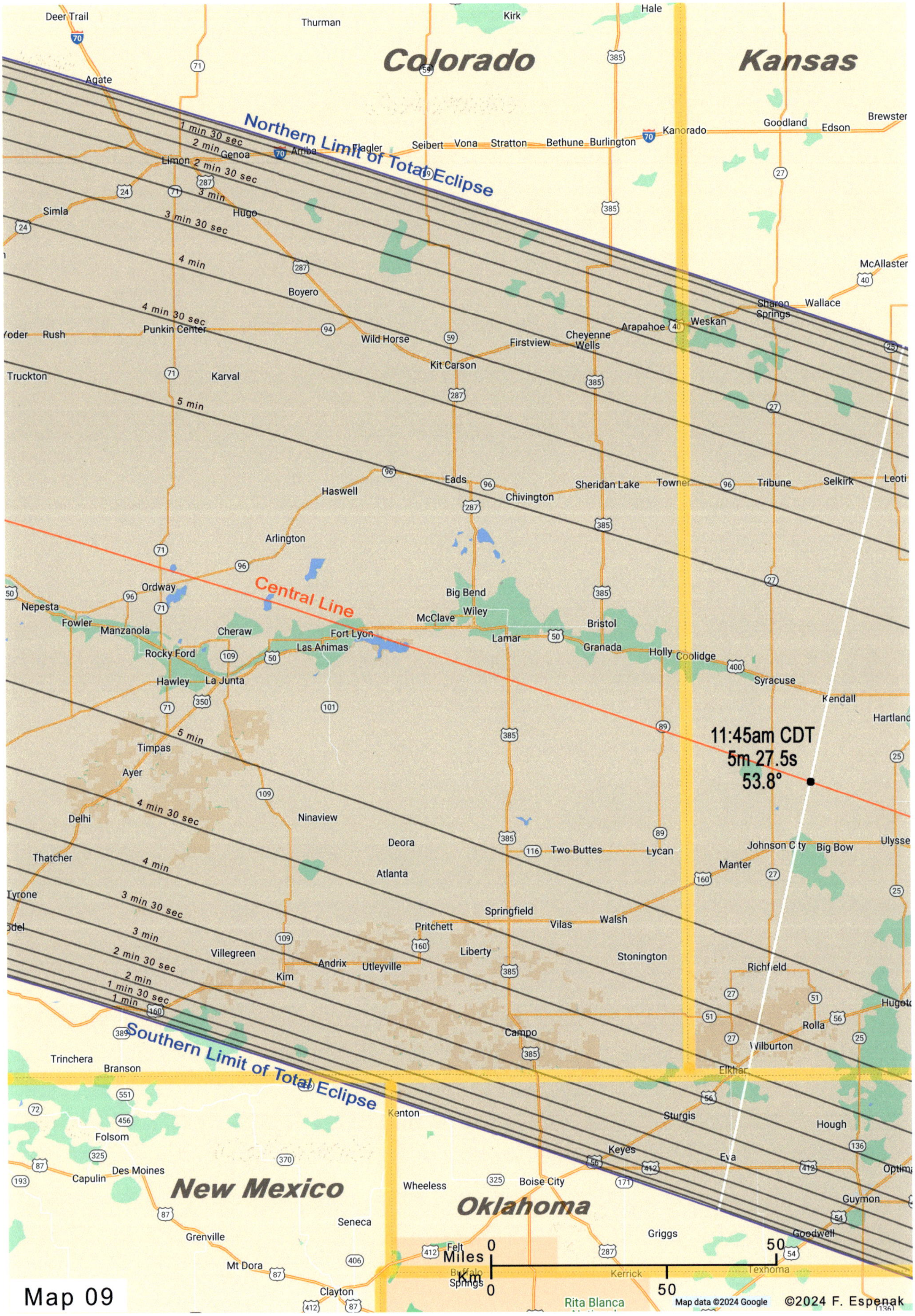
Colorado
Kansas
New Mexico
Oklahoma
Northern Limit of Total Eclipse
Southern Limit of Total Eclipse
Central Line
1 min 30 sec
2 min
2 min 30 sec
3 min
3 min 30 sec
4 min
4 min 30 sec
5 min
11:45am CDT
5m 27.5s
53.8°
Miles 0 50
Km 0 50
Map 09
Map data ©2024 Google
©2024 F. Espenak
Deer Trail
Agate
Thurman
Kirk
Hale
Limon
Genoa
Arriba
Flagler
Seibert
Vona
Stratton
Bethune
Burlington
Kanorado
Goodland
Edson
Brewster
Simla
Hugo
Boyero
Wild Horse
Kit Carson
Firstview
Cheyenne Wells
Arapahoe
Weskan
Sharon Springs
Wallace
McAllaster
Rush
Punkin Center
Truckton
Karval
Haswell
Eads
Chivington
Sheridan Lake
Towner
Tribune
Selkirk
Leoti
Arlington
Ordway
Nepesta
Fowler
Manzanola
Cheraw
Rocky Ford
Hawley
La Junta
Las Animas
Fort Lyon
McClave
Big Bend
Wiley
Lamar
Bristol
Granada
Holly
Coolidge
Syracuse
Kendall
Hartland
Timpas
Ayer
Delhi
Thatcher
Tyrone
Ninaview
Deora
Atlanta
Two Buttes
Lycan
Johnson City
Big Bow
Ulysses
Manter
Springfield
Pritchett
Vilas
Walsh
Villegreen
Andrix
Utleyville
Liberty
Stonington
Kim
Richfield
Rolla
Hugoton
Campo
Wilburton
Elkhart
Trinchera
Branson
Kenton
Sturgis
Hough
Folsom
Keyes
Eva
Capulin
Des Moines
Wheeless
Boise City
Optima
Guymon
Seneca
Griggs
Goodwell
Grenville
Felt
Buffalo Springs
Kerrick
Texhoma
Mt Dora
Clayton
Rita Blanca

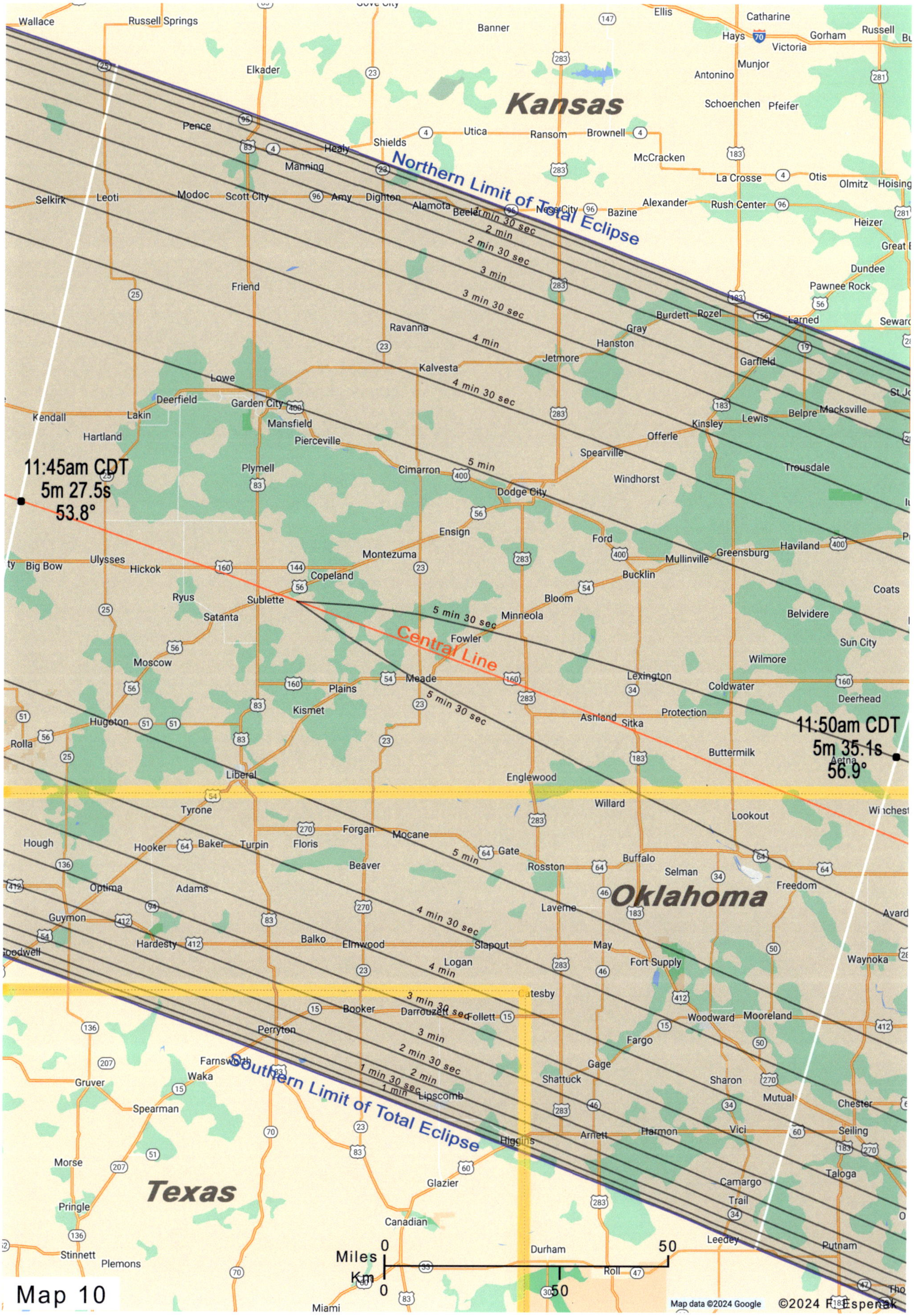
Kansas
Oklahoma
Texas
Northern Limit of Total Eclipse
Southern Limit of Total Eclipse
Central Line
1 min 30 sec
2 min
2 min 30 sec
3 min
3 min 30 sec
4 min
4 min 30 sec
5 min
5 min 30 sec
11:45am CDT
5m 27.5s
53.8°
11:50am CDT
5m 35.1s
56.9°
Miles 0 50
Km 0 50
Map 10
Map data ©2024 Google
©2024 F. Espenak
Wallace
Russell Springs
Elkader
Banner
Ellis
Catharine
Hays
Gorham
Russell
Victoria
Antonino
Munjor
Schoenchen
Pfeifer
Pence
Healy
Shields
Utica
Ransom
Brownell
McCracken
La Crosse
Otis
Olmitz
Selkirk
Leoti
Modoc
Scott City
Amy
Dighton
Alamota
Beeler
Ness City
Bazine
Alexander
Rush Center
Heizer
Dundee
Pawnee Rock
Friend
Ravanna
Burdett
Rozel
Larned
Gray
Hanston
Jetmore
Garfield
Kalvesta
Lowe
Deerfield
Garden City
Mansfield
Kendall
Lakin
Hartland
Pierceville
Kinsley
Lewis
Belpre
Macksville
Offerle
Spearville
Trousdale
Plymell
Cimarron
Windhorst
Dodge City
Ensign
Ford
Montezuma
Haviland
Greensburg
Mullinville
Big Bow
Ulysses
Hickok
Copeland
Bucklin
Coats
Ryus
Sublette
Satanta
Bloom
Minneola
Belvidere
Fowler
Moscow
Sun City
Wilmore
Meade
Lexington
Coldwater
Deerhead
Plains
Kismet
Hugoton
Ashland
Sitka
Protection
Rolla
Liberal
Buttermilk
Englewood
Tyrone
Willard
Lookout
Hough
Hooker
Baker
Turpin
Forgan
Floris
Mocane
Gate
Beaver
Rosston
Buffalo
Selman
Freedom
Optima
Adams
Laverne
Guymon
Hardesty
Balko
Elmwood
Slapout
May
Fort Supply
Waynoka
Logan
Catesby
Booker
Darrouzett
Follett
Woodward
Mooreland
Perryton
Fargo
Farnsworth
Waka
Gruver
Gage
Shattuck
Sharon
Mutual
Chester
Spearman
Lipscomb
Arnett
Harmon
Vici
Seiling
Higgins
Morse
Glazier
Taloga
Camargo
Trail
Pringle
Canadian
Leedey
Stinnett
Plemons
Durham
Putnam
Roll
Miami

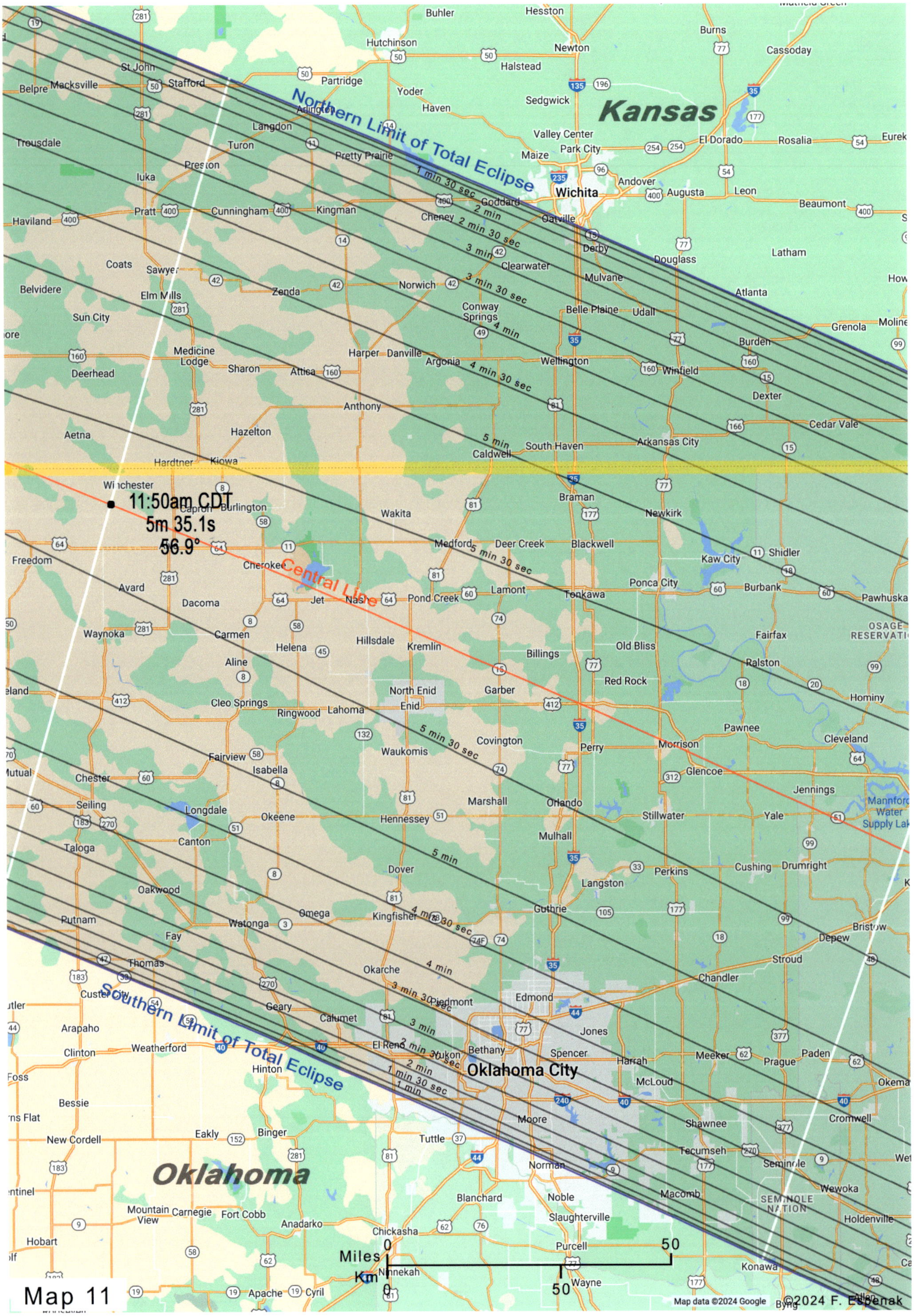
Kansas
Oklahoma
Northern Limit of Total Eclipse
Southern Limit of Total Eclipse
Central Line
11:50am CDT
5m 35.1s
56.9°
1 min 30 sec
2 min
2 min 30 sec
3 min
3 min 30 sec
4 min
4 min 30 sec
5 min
5 min 30 sec
5 min 30 sec
5 min
4 min
4 min 30 sec
3 min 30 sec
3 min
2 min 30 sec
2 min
1 min 30 sec
1 min
Hutchinson
Newton
Wichita
El Dorado
Pratt
Kingman
Wellington
Winfield
Arkansas City
Medicine Lodge
Kiowa
Caldwell
Enid
Ponca City
Stillwater
Perry
Guthrie
Edmond
Oklahoma City
Norman
Moore
Shawnee
Kingfisher
El Reno
Weatherford
Clinton
Chickasha
Anadarko
Woodward
Alva
Cherokee
Medford
Blackwell
Tonkawa
Pawnee
Cushing
Stroud
Chandler
Seminole
Miles
Km
0
50
Map 11
Map data ©2024 Google
©2024 F. Espenak

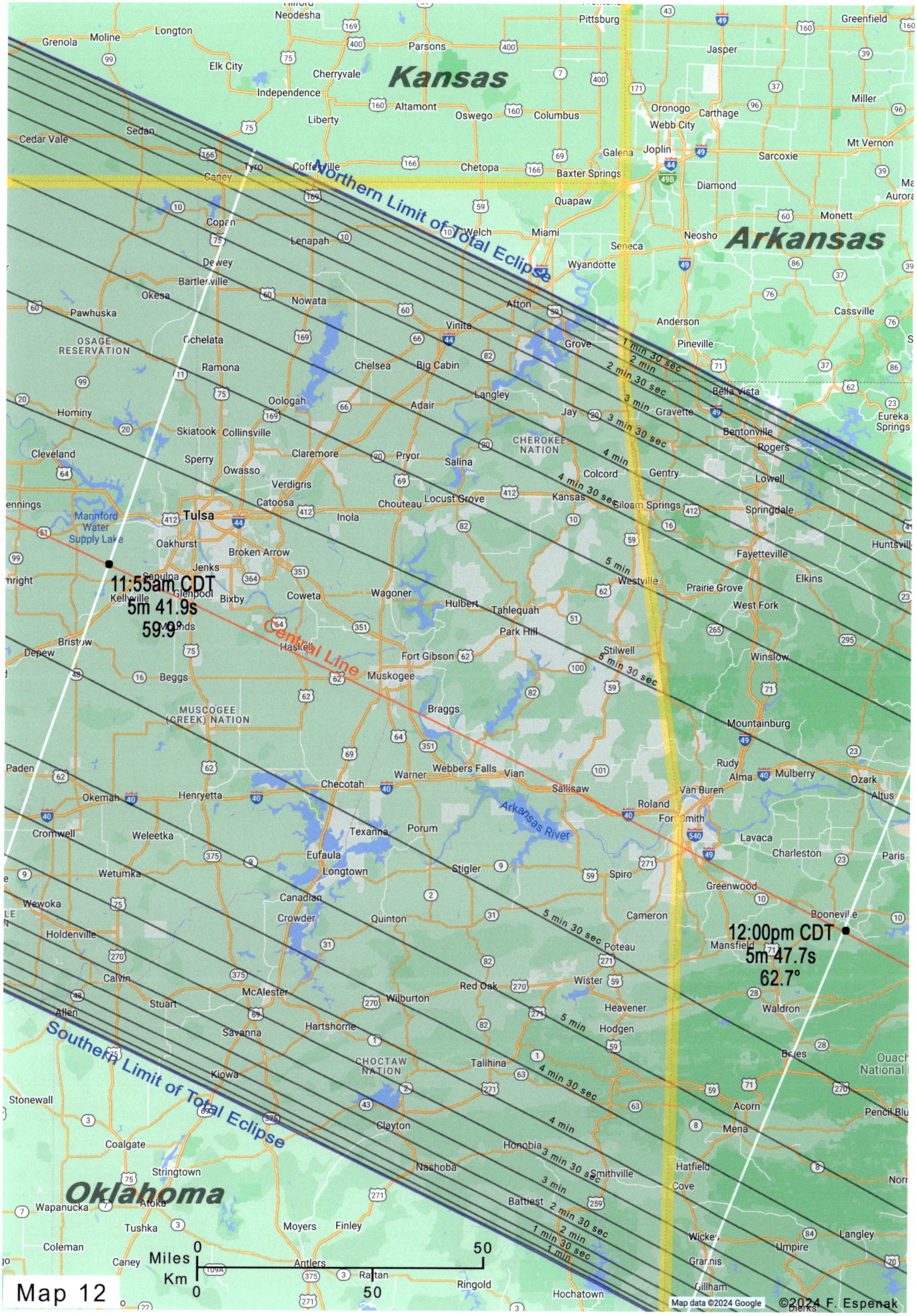
Kansas
Arkansas
Oklahoma
Northern Limit of Total Eclipse
Central Line
Southern Limit of Total Eclipse
11:55am CDT
5m 41.9s
59.9°
12:00pm CDT
5m 47.7s
62.7°
1 min 30 sec
2 min
2 min 30 sec
3 min
3 min 30 sec
4 min
4 min 30 sec
5 min
5 min 30 sec
1 min
OSAGE RESERVATION
CHEROKEE NATION
MUSCOGEE (CREEK) NATION
CHOCTAW NATION
Arkansas River
Tulsa
Muskogee
Fort Smith
Joplin
Fayetteville
Bartlesville
McAlester
Miles
Km
0
50
Map 12
Map data ©2024 Google
©2024 F. Espenak

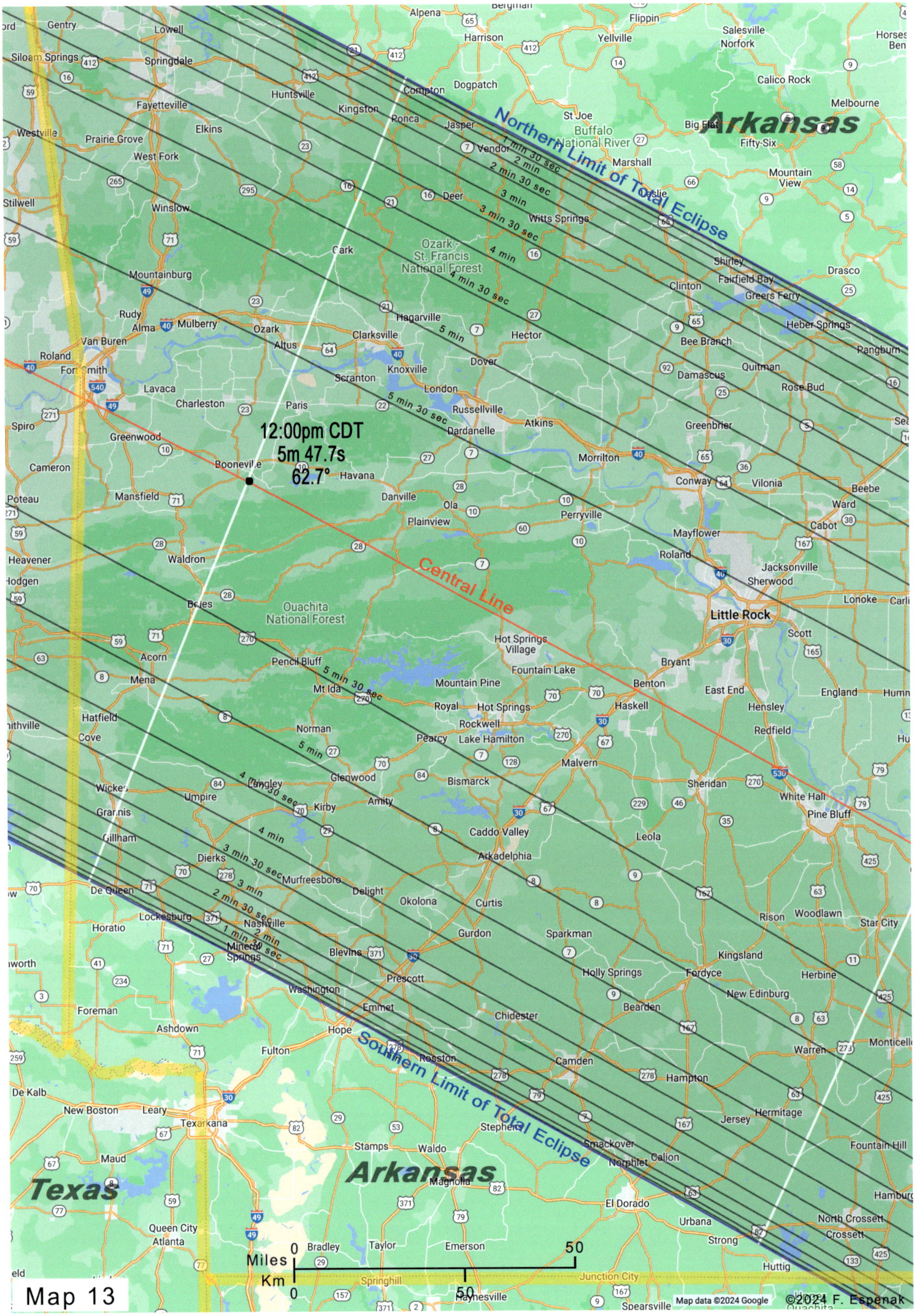
Northern Limit of Total Eclipse
1 min 30 sec
2 min
2 min 30 sec
3 min
3 min 30 sec
4 min
4 min 30 sec
5 min
5 min 30 sec
12:00pm CDT
5m 47.7s
62.7°
Central Line
5 min 30 sec
5 min
4 min 30 sec
4 min
3 min 30 sec
3 min
2 min 30 sec
2 min
1 min 30 sec
Southern Limit of Total Eclipse
Arkansas
Arkansas
Texas
Harrison
Yellville
Flippin
Salesville
Norfork
Calico Rock
Melbourne
Alpena
Gentry
Lowell
Siloam Springs
Springdale
Huntsville
Kingston
Compton
Dogpatch
Fayetteville
Westville
Prairie Grove
West Fork
Elkins
Ponca
Jasper
Vendor
St Joe
Buffalo National River
Marshall
Big Flat
Fifty-Six
Mountain View
Stilwell
Winslow
Deer
Witts Springs
Ozark - St. Francis National Forest
Oark
Clinton
Shirley
Fairfield Bay
Greers Ferry
Drasco
Heber Springs
Mountainburg
Rudy
Alma
Mulberry
Ozark
Altus
Clarksville
Hagarville
Hector
Bee Branch
Van Buren
Roland
Fort Smith
Lavaca
Charleston
Scranton
Knoxville
Dover
London
Damascus
Quitman
Pangburn
Rose Bud
Spiro
Greenwood
Paris
Russellville
Dardanelle
Atkins
Greenbrier
Cameron
Booneville
Havana
Morrilton
Conway
Vilonia
Poteau
Mansfield
Danville
Ola
Plainview
Perryville
Beebe
Ward
Cabot
Mayflower
Heavener
Hodgen
Waldron
Roland
Jacksonville
Sherwood
Lonoke
Little Rock
Ouachita National Forest
Bryant
Scott
Hot Springs Village
Acorn
Mena
Pencil Bluff
Mt Ida
Mountain Pine
Fountain Lake
Benton
East End
England
Royal
Hot Springs
Haskell
Hensley
Redfield
Hatfield
Cove
Norman
Rockwell
Pearcy
Lake Hamilton
Malvern
Glenwood
Bismarck
Sheridan
White Hall
Pine Bluff
Umpire
Kirby
Amity
Caddo Valley
Leola
Gillham
Dierks
Arkadelphia
Murfreesboro
Delight
De Queen
Okolona
Curtis
Rison
Woodlawn
Star City
Horatio
Lockesburg
Nashville
Gurdon
Sparkman
Mineral Springs
Blevins
Kingsland
Holly Springs
Fordyce
Herbine
Prescott
New Edinburg
Washington
Foreman
Bearden
Emmet
Chidester
Ashdown
Hope
Monticello
Warren
Fulton
Rosston
Camden
De Kalb
Hampton
New Boston
Leary
Texarkana
Hermitage
Jersey
Stephens
Stamps
Waldo
Smackover
Norphlet
Calion
Fountain Hill
Maud
Magnolia
El Dorado
Hamburg
Urbana
Queen City
Atlanta
North Crossett
Crossett
Bradley
Taylor
Emerson
Strong
Huttig
Springhill
Junction City
Haynesville
Spearsville
Miles
Km
0
50
0
50
Map data ©2024 Google
©2024 F. Espenak

Map 13

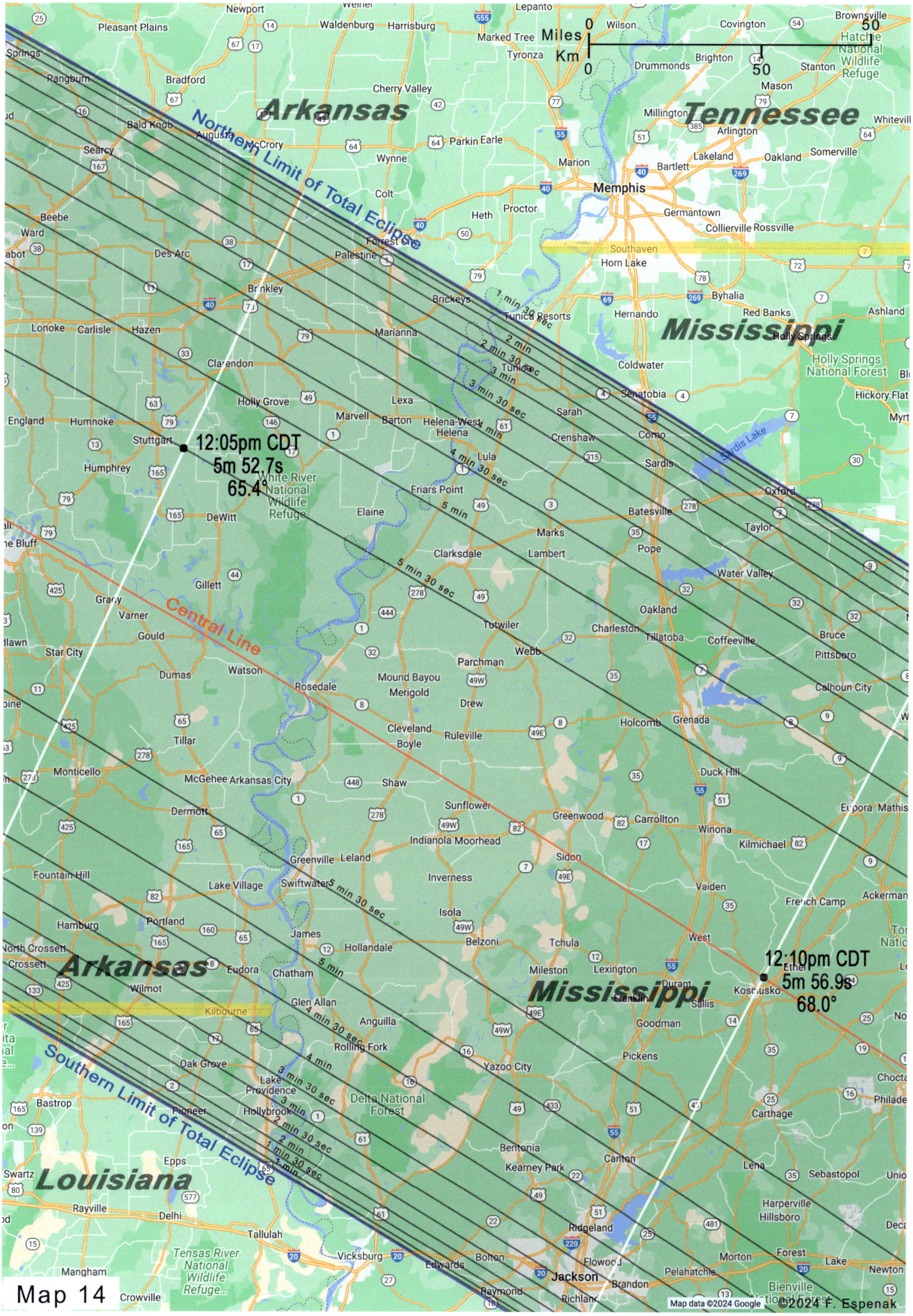
Miles 0 50
Km 0 50
Arkansas
Tennessee
Mississippi
Arkansas
Mississippi
Louisiana
Northern Limit of Total Eclipse
Central Line
Southern Limit of Total Eclipse
1 min 30 sec
2 min
2 min 30 sec
3 min
3 min 30 sec
4 min
4 min 30 sec
5 min
5 min 30 sec
5 min 30 sec
5 min
4 min 30 sec
4 min
3 min 30 sec
3 min
2 min 30 sec
2 min
1 min 30 sec
1 min
12:05pm CDT
5m 52.7s
65.4°
12:10pm CDT
5m 56.9s
68.0°
Memphis
Jackson
Map 14
Map data ©2024 Google
©2024 F. Espenak

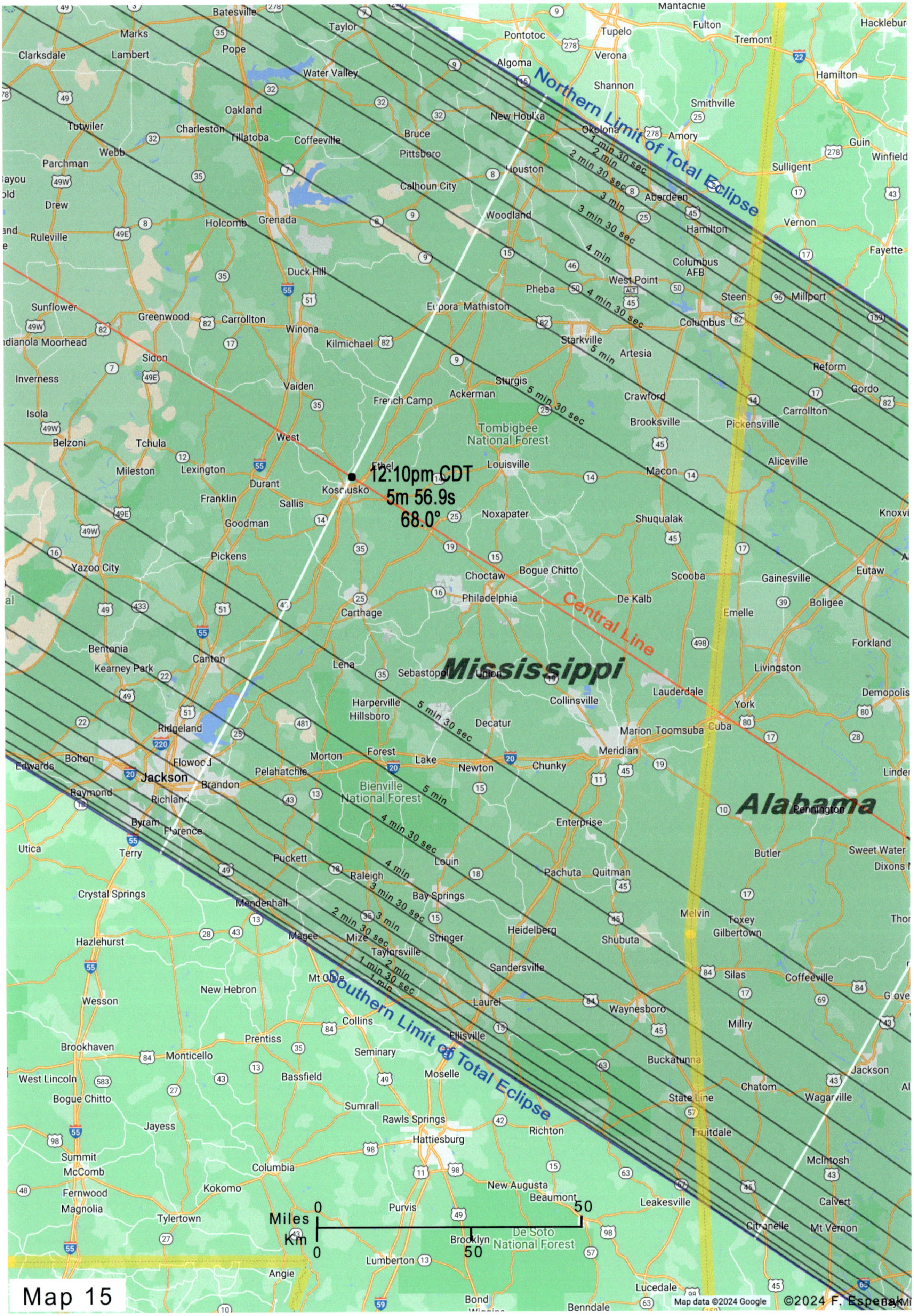
Northern Limit of Total Eclipse
Southern Limit of Total Eclipse
Central Line
12:10pm CDT
5m 56.9s
68.0°
1 min
1 min 30 sec
2 min
2 min 30 sec
3 min
3 min 30 sec
4 min
4 min 30 sec
5 min
5 min 30 sec
Mississippi
Alabama
Miles 0 50
Km 0 50
Map 15
Map data ©2024 Google
©2024 F. Espenak

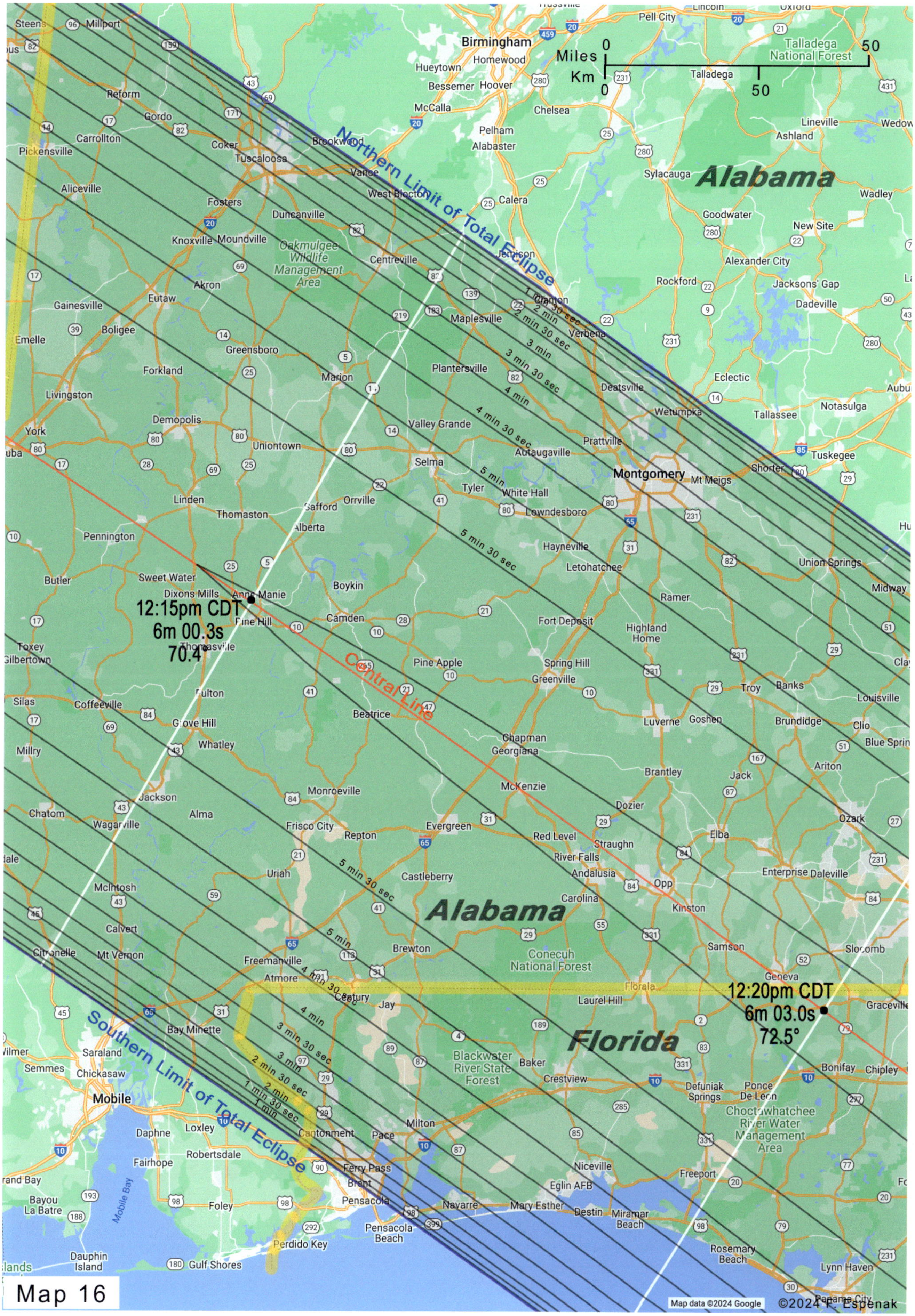
Map 16
Alabama
Florida
Northern Limit of Total Eclipse
Southern Limit of Total Eclipse
Central Line
12:15pm CDT
6m 00.3s
70.4°
12:20pm CDT
6m 03.0s
72.5°
1 min
1 min 30 sec
2 min
2 min 30 sec
3 min
3 min 30 sec
4 min
4 min 30 sec
5 min
5 min 30 sec
Miles 0 50
Km 0 50
Birmingham
Tuscaloosa
Montgomery
Selma
Mobile
Pensacola
Talladega National Forest
Oakmulgee Wildlife Management Area
Conecuh National Forest
Blackwater River State Forest
Choctawhatchee River Water Management Area
Mobile Bay
Map data ©2024 Google
©2024 F. Espenak

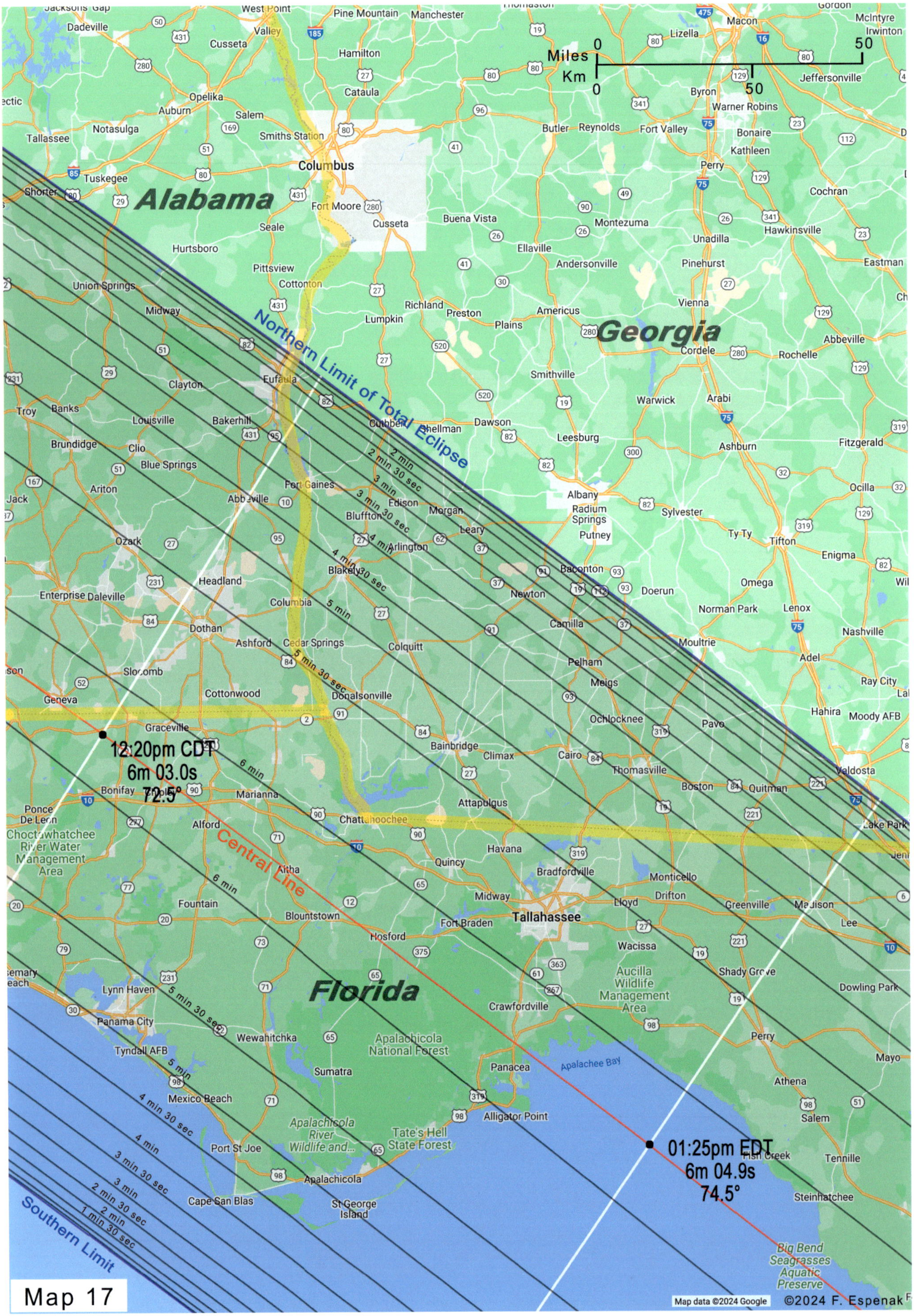
Miles 0 50
Km 0 50
Alabama
Georgia
Florida
Northern Limit of Total Eclipse
2 min
2 min 30 sec
3 min
3 min 30 sec
4 min
4 min 30 sec
5 min
5 min 30 sec
6 min
Central Line
6 min
5 min 30 sec
5 min
4 min 30 sec
4 min
3 min 30 sec
3 min
2 min 30 sec
2 min
1 min 30 sec
Southern Limit
12:20pm CDT
6m 03.0s
72.5°
01:25pm EDT
6m 04.9s
74.5°
Columbus
Dothan
Albany
Tallahassee
Macon
Valdosta
Panama City
Map 17
Map data ©2024 Google
©2024 F. Espenak

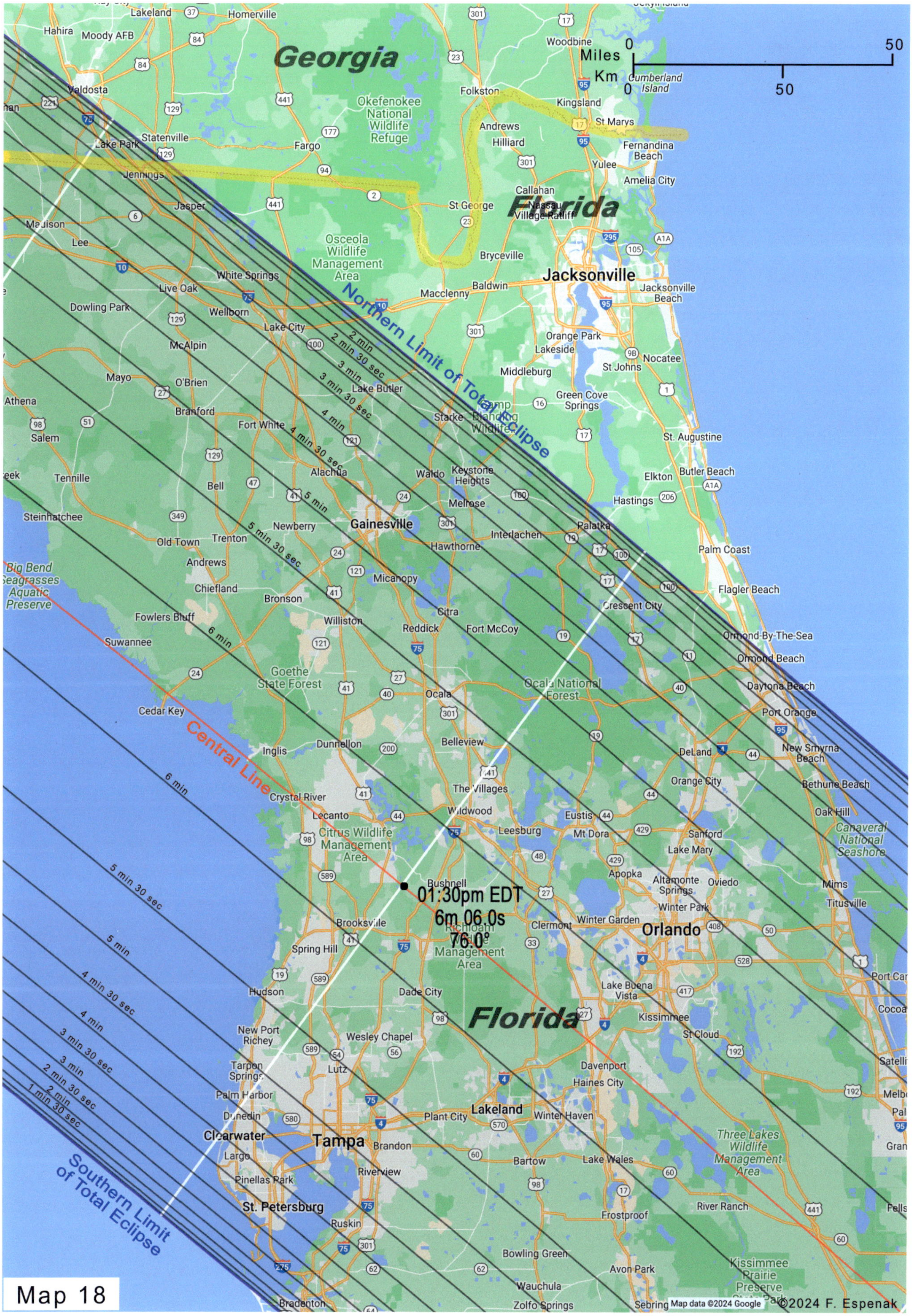
Georgia
Florida
Miles 0 50
Km 0 50
Northern Limit of Total Eclipse
Southern Limit of Total Eclipse
Central Line
2 min
2 min 30 sec
3 min
3 min 30 sec
4 min
4 min 30 sec
5 min
5 min 30 sec
6 min
6 min
5 min 30 sec
5 min
4 min 30 sec
4 min
3 min 30 sec
3 min
2 min 30 sec
2 min
1 min 30 sec
01:30pm EDT
6m 06.0s
76.0°
Valdosta
Lake Park
Statenville
Jennings
Jasper
Madison
Lee
White Springs
Live Oak
Dowling Park
Wellborn
Lake City
McAlpin
Mayo
O'Brien
Branford
Athena
Salem
Fort White
Lake Butler
Starke
Alachua
Tennille
Bell
Steinhatchee
Old Town
Trenton
Newberry
Gainesville
Andrews
Chiefland
Bronson
Fowlers Bluff
Suwannee
Cedar Key
Inglis
Dunnellon
Crystal River
Lecanto
Williston
Micanopy
Citra
Reddick
Fort McCoy
Ocala
Belleview
The Villages
Wildwood
Leesburg
Eustis
Mt Dora
Bushnell
Brooksville
Spring Hill
Hudson
Dade City
New Port Richey
Wesley Chapel
Tarpon Springs
Lutz
Palm Harbor
Dunedin
Clearwater
Largo
Tampa
Brandon
Riverview
Pinellas Park
St. Petersburg
Ruskin
Bradenton
Plant City
Lakeland
Winter Haven
Bartow
Lake Wales
Frostproof
Bowling Green
Wauchula
Zolfo Springs
Avon Park
Sebring
Davenport
Haines City
Kissimmee
St Cloud
Clermont
Winter Garden
Orlando
Lake Buena Vista
Apopka
Altamonte Springs
Winter Park
Oviedo
Sanford
Lake Mary
Orange City
DeLand
Deltona
New Smyrna Beach
Bethune Beach
Oak Hill
Canaveral National Seashore
Mims
Titusville
Port Orange
Daytona Beach
Ormond Beach
Ormond-By-The-Sea
Flagler Beach
Palm Coast
Crescent City
Palatka
Interlachen
Hawthorne
Melrose
Keystone Heights
Waldo
Hastings
Elkton
Butler Beach
St. Augustine
Green Cove Springs
Middleburg
Orange Park
Lakeside
St Johns
Nocatee
Jacksonville
Jacksonville Beach
Baldwin
Macclenny
Bryceville
Callahan
St George
Hilliard
Andrews
Folkston
Kingsland
St Marys
Fernandina Beach
Yulee
Amelia City
Woodbine
Cumberland Island
Homerville
Lakeland
Hahira
Moody AFB
Fargo
Okefenokee National Wildlife Refuge
Osceola Wildlife Management Area
Goethe State Forest
Ocala National Forest
Citrus Wildlife Management Area
Richloam Management Area
Three Lakes Wildlife Management Area
Kissimmee Prairie Preserve
Big Bend Seagrasses Aquatic Preserve
River Ranch
Map 18
Map data ©2024 Google
©2024 F. Espenak

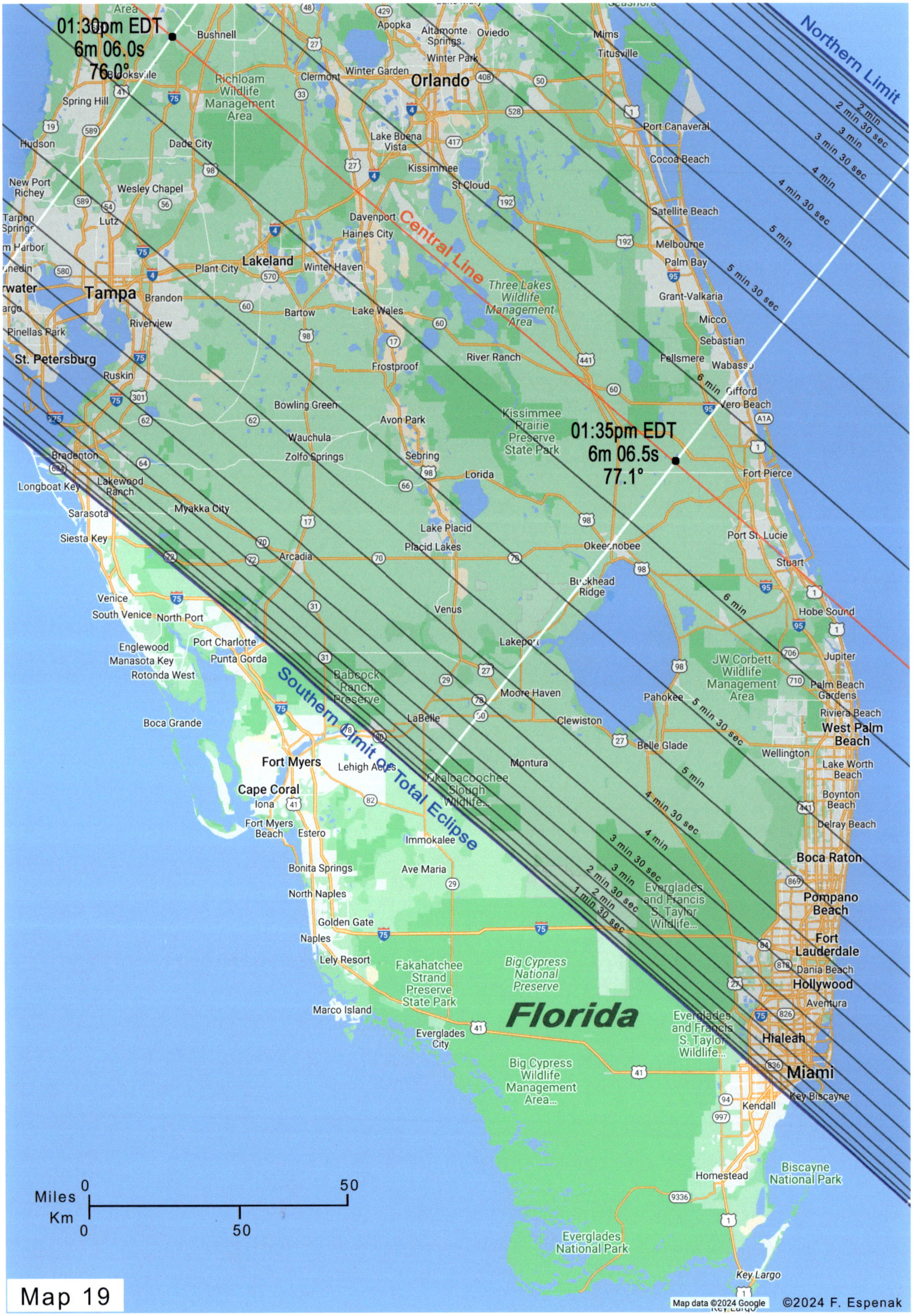
01:30pm EDT
6m 06.0s
76.0°
01:35pm EDT
6m 06.5s
77.1°
Northern Limit
Central Line
Southern Limit of Total Eclipse
2 min
2 min 30 sec
3 min
3 min 30 sec
4 min
4 min 30 sec
5 min
5 min 30 sec
6 min
1 min 30 sec
Florida
Orlando
Tampa
St. Petersburg
Lakeland
Fort Myers
Cape Coral
Miami
Hialeah
Fort Lauderdale
West Palm Beach
Boca Raton
Okeechobee
Miles
Km
0
50
Map data ©2024 Google
©2024 F. Espenak
Map 19

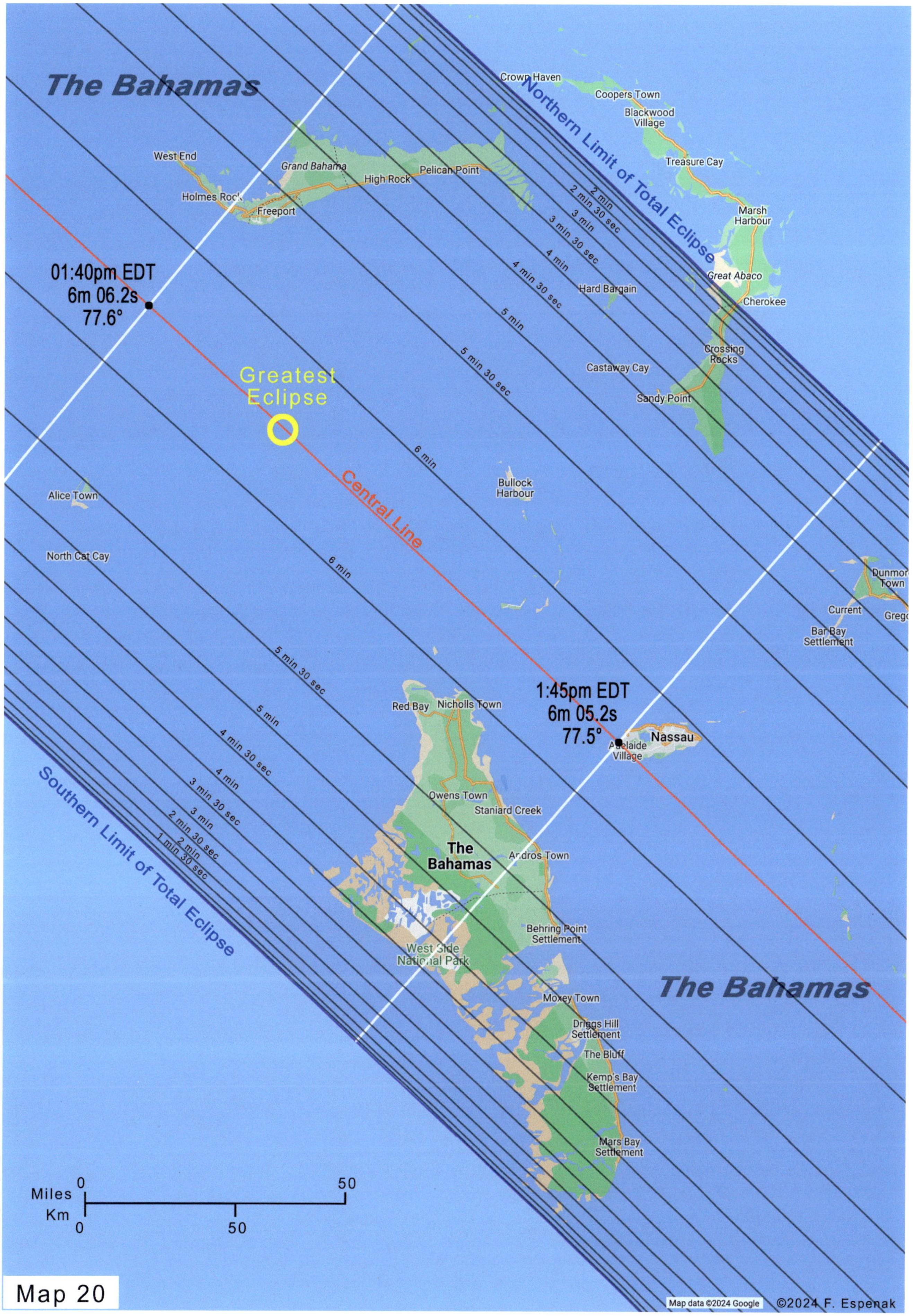
The Bahamas
Crown Haven
Coopers Town
Blackwood Village
Northern Limit of Total Eclipse
Treasure Cay
West End
Grand Bahama
Pelican Point
High Rock
Holmes Rock
Freeport
2 min
2 min 30 sec
3 min
3 min 30 sec
4 min
4 min 30 sec
5 min
5 min 30 sec
6 min
Marsh Harbour
Great Abaco
Cherokee
Hard Bargain
01:40pm EDT
6m 06.2s
77.6°
Crossing Rocks
Castaway Cay
Sandy Point
Greatest Eclipse
Bullock Harbour
Central Line
Alice Town
North Cat Cay
Dunmore Town
Current
Bar Bay Settlement
Red Bay
Nicholls Town
1:45pm EDT
6m 05.2s
77.5°
Nassau
Adelaide Village
Owens Town
Staniard Creek
The Bahamas
Andros Town
1 min 30 sec
Southern Limit of Total Eclipse
Behring Point Settlement
West Side National Park
Moxey Town
Driggs Hill Settlement
The Bluff
Kemp's Bay Settlement
Mars Bay Settlement
Miles 0 50
Km 0 50
Map 20
Map data ©2024 Google
©2024 F. Espenak

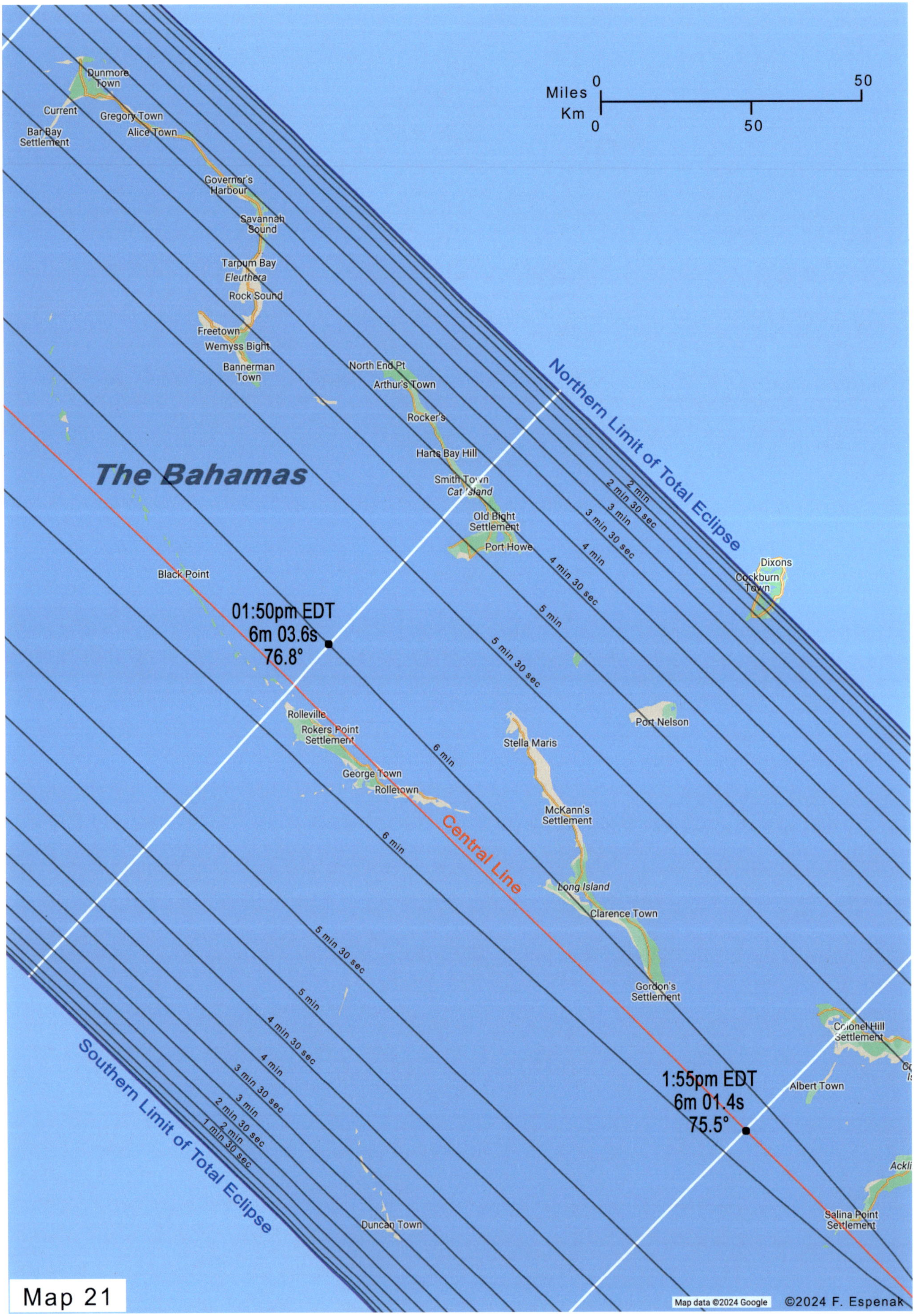
Dunmore Town
Current
Gregory Town
Alice Town
Bar Bay Settlement
Governor's Harbour
Savannah Sound
Tarpum Bay
Eleuthera
Rock Sound
Freetown
Wemyss Bight
Bannerman Town
North End Pt
Arthur's Town
Rocker's
Harts Bay Hill
Smith Town
Cat Island
Old Bight Settlement
Port Howe
Miles 0 50
Km 0 50
Northern Limit of Total Eclipse
2 min
2 min 30 sec
3 min
3 min 30 sec
4 min
4 min 30 sec
5 min
5 min 30 sec
6 min
Dixons
Cockburn Town
The Bahamas
Black Point
01:50pm EDT
6m 03.6s
76.8°
Port Nelson
Rolleville
Rokers Point Settlement
Stella Maris
George Town
Rolletown
McKann's Settlement
Central Line
6 min
Long Island
Clarence Town
5 min 30 sec
5 min
Gordon's Settlement
4 min 30 sec
4 min
3 min 30 sec
3 min
2 min 30 sec
2 min
1 min 30 sec
Southern Limit of Total Eclipse
Colonel Hill Settlement
1:55pm EDT
6m 01.4s
75.5°
Albert Town
Salina Point Settlement
Duncan Town
Map 21
Map data ©2024 Google
©2024 F. Espenak

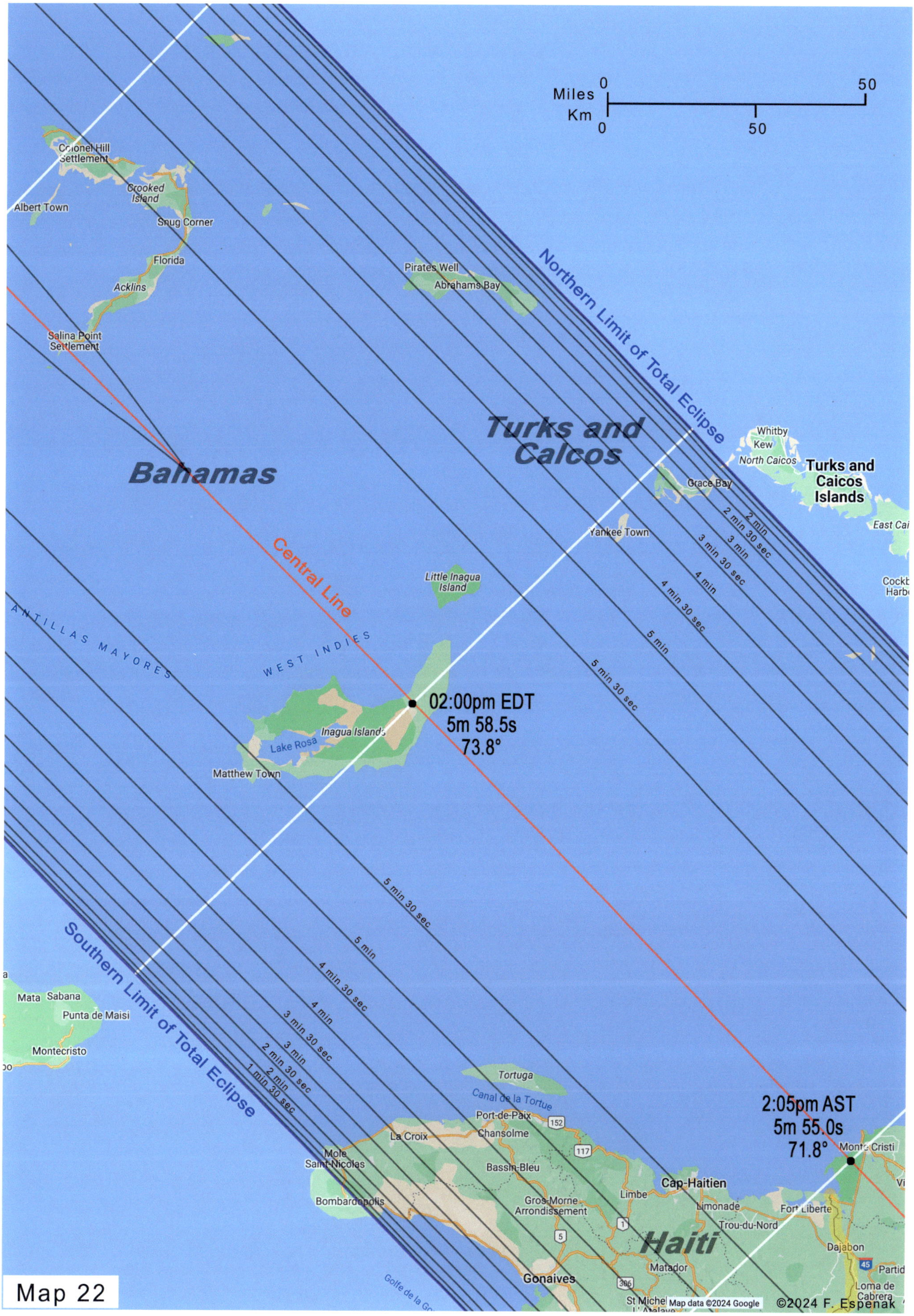
Miles
0
50
Km
0
50
Colonel Hill Settlement
Crooked Island
Albert Town
Snug Corner
Florida
Acklins
Salina Point Settlement
Pirates Well
Abrahams Bay
Northern Limit of Total Eclipse
Turks and Calcos
Bahamas
Whitby
Kew
North Caicos
Turks and Caicos Islands
Grace Bay
East Cai
Yankee Town
2 min
2 min 30 sec
3 min
3 min 30 sec
4 min
4 min 30 sec
5 min
5 min 30 sec
Central Line
Little Inagua Island
ANTILLAS MAYORES
WEST INDIES
02:00pm EDT
5m 58.5s
73.8°
Inagua Islands
Lake Rosa
Matthew Town
5 min 30 sec
5 min
4 min 30 sec
4 min
3 min 30 sec
3 min
2 min 30 sec
2 min
1 min 30 sec
Southern Limit of Total Eclipse
Mata
Sabana
Punta de Maisi
Montecristo
Tortuga
Canal de la Tortue
Port-de-Paix
Chansolme
La Croix
Mole Saint-Nicolas
Bassin-Bleu
Bombardopolis
Gros-Morne Arrondissement
Limbe
Cap-Haitien
Limonade
Trou-du-Nord
Fort Liberte
2:05pm AST
5m 55.0s
71.8°
Monte Cristi
Haiti
Dajabon
Matador
Gonaives
St Michel
Map data ©2024 Google
©2024 F. Espenak
Map 22

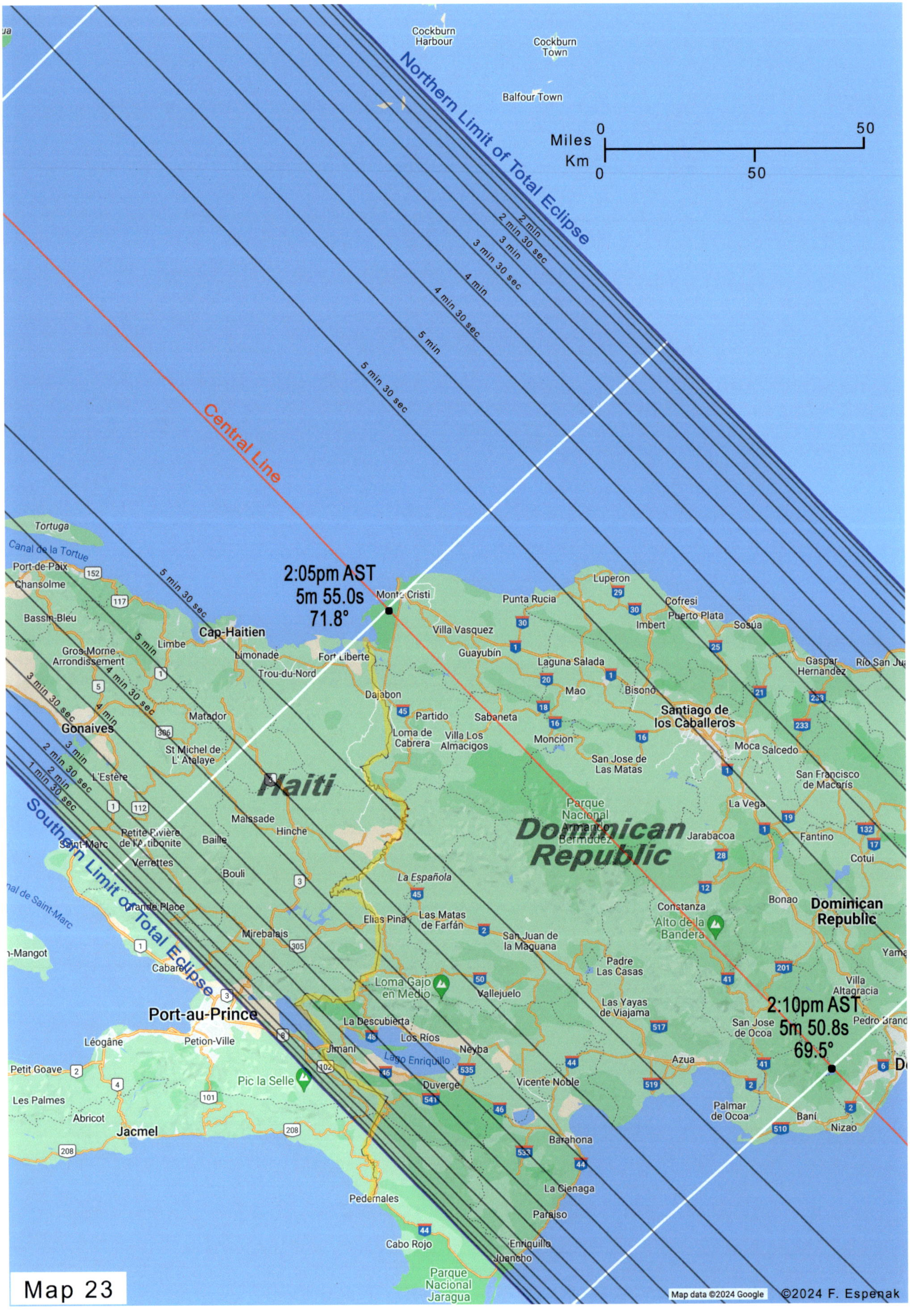
Northern Limit of Total Eclipse
Southern Limit of Total Eclipse
Central Line
2 min
2 min 30 sec
3 min
3 min 30 sec
4 min
4 min 30 sec
5 min
5 min 30 sec
1 min 30 sec
2:05pm AST
5m 55.0s
71.8°
2:10pm AST
5m 50.8s
69.5°
Miles 0 50
Km 0 50
Haiti
Dominican Republic
Cockburn Harbour
Cockburn Town
Balfour Town
Tortuga
Cap-Haitien
Gonaives
Port-au-Prince
Jacmel
Monte Cristi
Santiago de los Caballeros
Puerto Plata
La Vega
Constanza
Pedernales
Barahona
Map 23
Map data ©2024 Google
©2024 F. Espenak

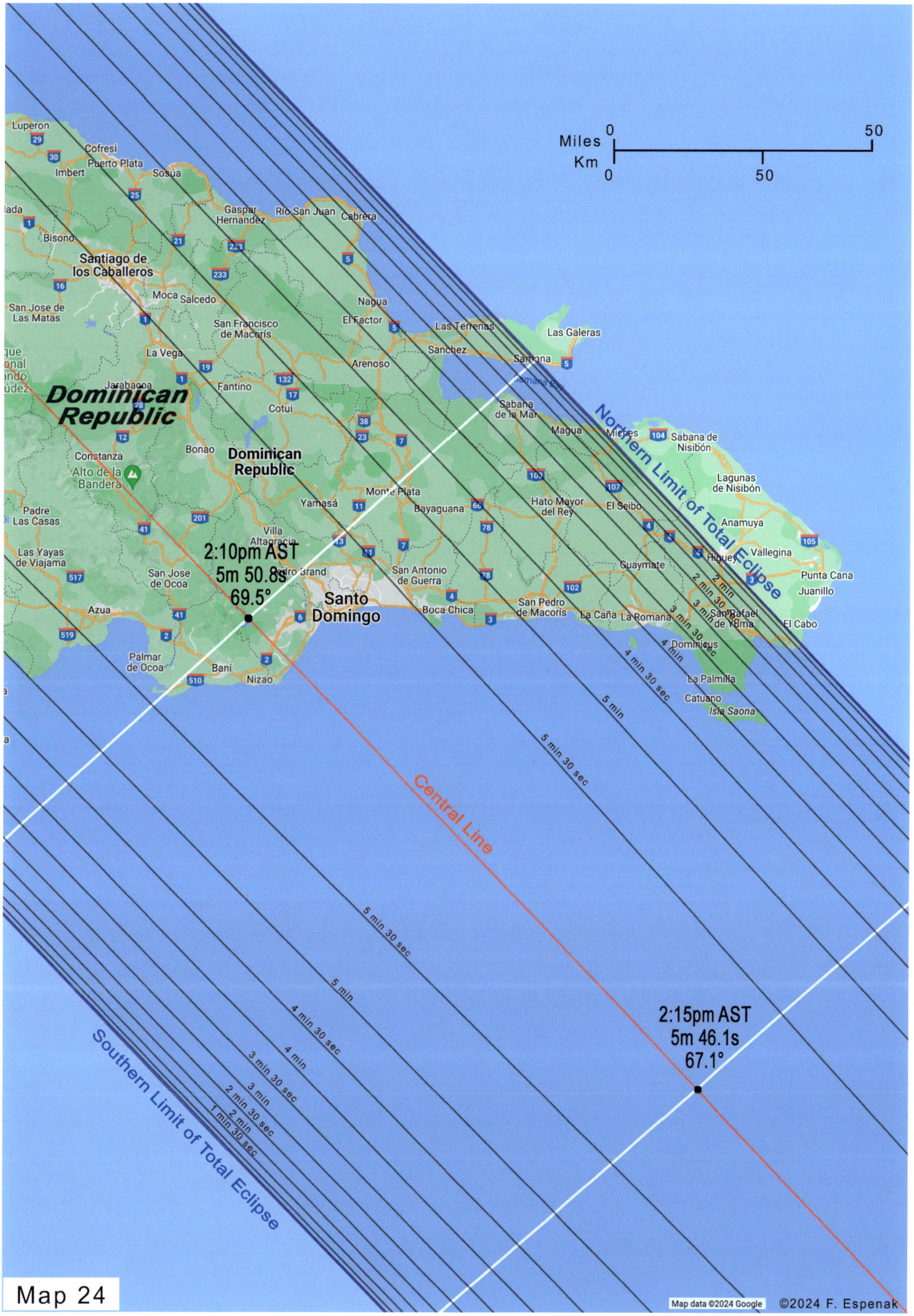
Map 24
Miles 0 50
Km 0 50
Dominican Republic
Santo Domingo
Northern Limit of Total Eclipse
Southern Limit of Total Eclipse
Central Line
2:10pm AST
5m 50.8s
69.5°
2:15pm AST
5m 46.1s
67.1°
5 min 30 sec
5 min
4 min 30 sec
4 min
3 min 30 sec
3 min
2 min 30 sec
2 min
1 min 30 sec
Map data ©2024 Google
©2024 F. Espenak

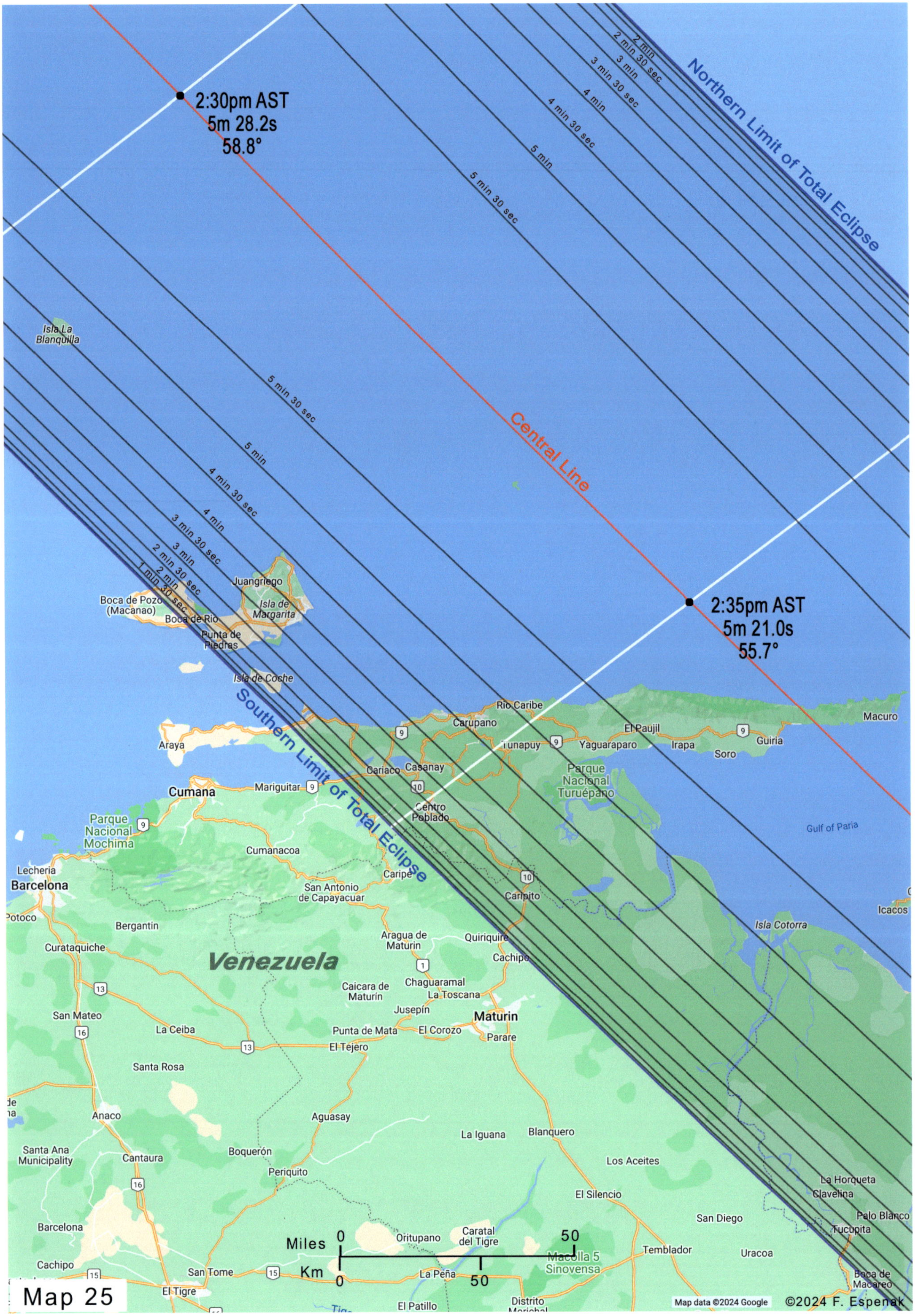
2:30pm AST
5m 28.2s
58.8°
2:35pm AST
5m 21.0s
55.7°
Northern Limit of Total Eclipse
Southern Limit of Total Eclipse
Central Line
2 min
2 min 30 sec
3 min
3 min 30 sec
4 min
4 min 30 sec
5 min
5 min 30 sec
1 min 30 sec
Isla La Blanquilla
Juangriego
Isla de Margarita
Boca de Pozo (Macanao)
Boca de Rio
Punta de Piedras
Isla de Coche
Rio Caribe
Carupano
El Paujil
Macuro
Tunapuy
Yaguaraparo
Irapa
Soro
Guiria
Araya
Cariaco
Casanay
Mariguitar
Cumana
Parque Nacional Turuépano
Centro Poblado
Gulf of Paria
Parque Nacional Mochima
Cumanacoa
Lecheria
Barcelona
Caripe
San Antonio de Capayacuar
Caripito
Icacos
Isla Cotorra
Bergantin
Aragua de Maturin
Quiriquire
Curataquiche
Cachipo
Venezuela
Caicara de Maturín
Chaguaramal
La Toscana
San Mateo
Jusepín
Maturin
La Ceiba
Punta de Mata
El Corozo
Parare
El Tejero
Santa Rosa
Anaco
Aguasay
La Iguana
Blanquero
Santa Ana Municipality
Cantaura
Boquerón
Los Aceites
Periquito
La Horqueta
Clavelina
El Silencio
San Diego
Palo Blanco
Tucupita
Barcelona
Caratal del Tigre
Oritupano
Miles 0 50
Km 0 50
Temblador
Uracoa
Macolla 5 Sinovensa
Cachipo
San Tome
La Peña
Boca de Macareo
El Tigre
El Patillo
Map 25
Map data ©2024 Google
©2024 F. Espenak

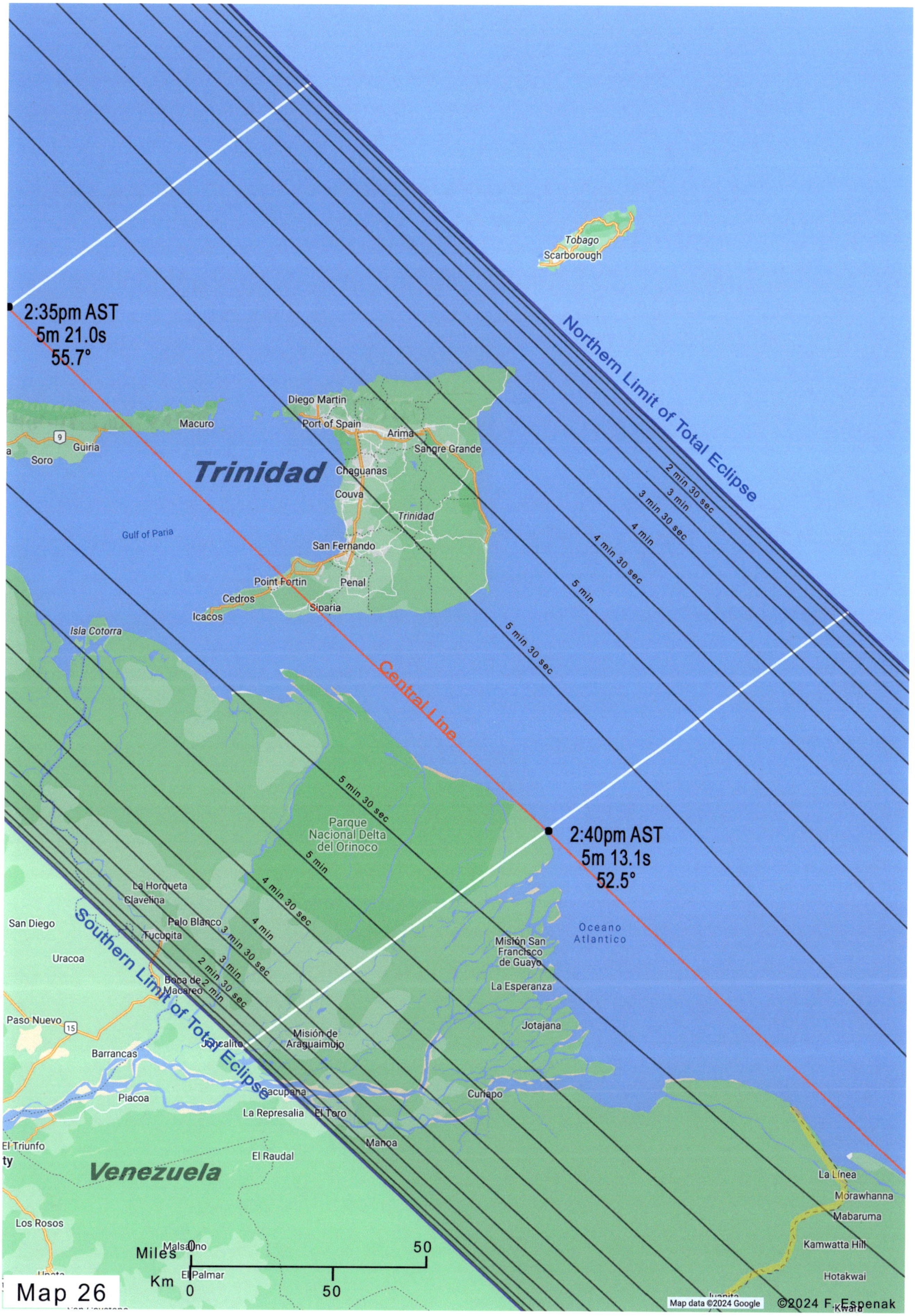
2:35pm AST
5m 21.0s
55.7°
2:40pm AST
5m 13.1s
52.5°
Northern Limit of Total Eclipse
Southern Limit of Total Eclipse
Central Line
2 min 30 sec
3 min
3 min 30 sec
4 min
4 min 30 sec
5 min
5 min 30 sec
Tobago
Scarborough
Trinidad
Diego Martin
Port of Spain
Arima
Sangre Grande
Chaguanas
Couva
San Fernando
Point Fortin
Penal
Cedros
Siparia
Icacos
Macuro
Guiria
Soro
Gulf of Paria
Isla Cotorra
Parque Nacional Delta del Orinoco
Oceano Atlantico
La Horqueta
Clavelina
San Diego
Palo Blanco
Tucupita
Uracoa
Boca de Macareo
Paso Nuevo
Barrancas
Misión San Francisco de Guayo
La Esperanza
Jotajana
Misión de Araguaimujo
Piacoa
Sacupana
La Represalia
El Toro
Curiapo
Manoa
El Triunfo
El Raudal
Venezuela
Los Rosos
La Línea
Morawhanna
Mabaruma
Kamwatta Hill
Hotakwai
Miles 0 50
Km 0 50
El Palmar
Map 26
Map data ©2024 Google
©2024 F. Espenak

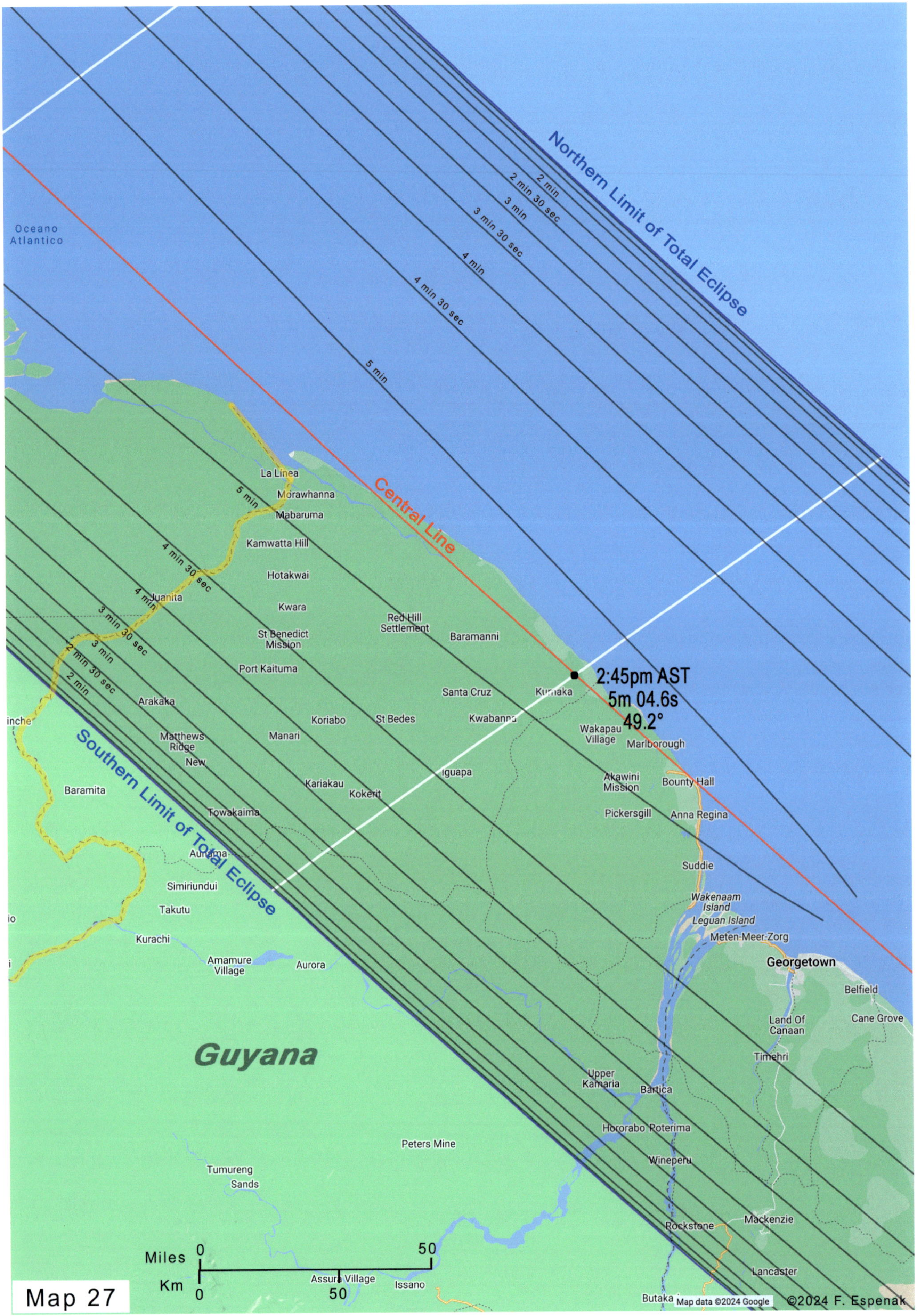

Northern Limit of Total Eclipse
2 min
2 min 30 sec
3 min
3 min 30 sec
4 min
4 min 30 sec
5 min
Oceano Atlantico
Central Line
La Linea
Morawhanna
Mabaruma
Kamwatta Hill
Hotakwai
Juanita
Kwara
Red Hill Settlement
Baramanni
St Benedict Mission
Port Kaituma
2:45pm AST
5m 04.6s
49.2°
Santa Cruz
Kumaka
Arakaka
Koriabo
St Bedes
Kwabanna
Wakapau Village
Marlborough
Matthews Ridge
New
Manari
Iguapa
Akawini Mission
Bounty Hall
Baramita
Kariakau
Kokerit
Pickersgill
Anna Regina
Towakaima
Southern Limit of Total Eclipse
Aurama
Suddie
Simiriundui
Takutu
Wakenaam Island
Leguan Island
Kurachi
Meten-Meer-Zorg
Amamure Village
Aurora
Georgetown
Belfield
Land Of Canaan
Cane Grove
Guyana
Timehri
Upper Kamaria
Bartica
Hororabo
Poterima
Peters Mine
Wineperu
Tumureng Sands
Rockstone
Mackenzie
Miles
0
50
Km
0
50
Lancaster
Assura Village
Issano
Map 27
Butaka
Map data ©2024 Google
©2024 F. Espenak

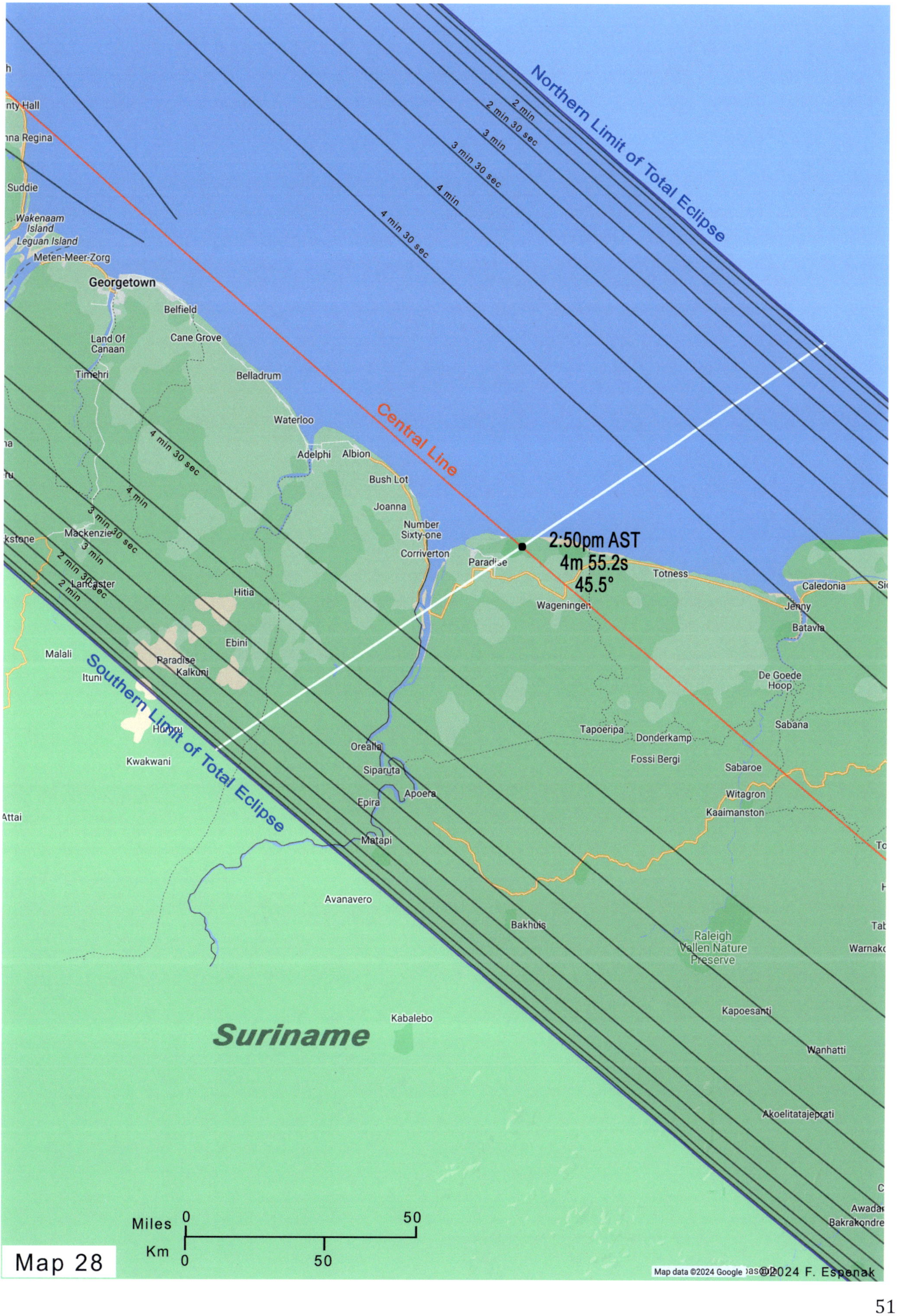
Northern Limit of Total Eclipse
2 min
2 min 30 sec
3 min
3 min 30 sec
4 min
4 min 30 sec
Central Line
2:50pm AST
4m 55.2s
45.5°
Southern Limit of Total Eclipse
Georgetown
Belfield
Cane Grove
Land Of Canaan
Timehri
Belladrum
Waterloo
Adelphi
Albion
Bush Lot
Joanna
Number Sixty-one
Corriverton
Paradise
Wageningen
Totness
Caledonia
Jenny
Batavia
De Goede Hoop
Sabana
Tapoeripa
Donderkamp
Fossi Bergi
Sabaroe
Witagron
Kaaimanston
Suddie
Wakenaam Island
Leguan Island
Meten-Meer-Zorg
Mackenzie
Lancaster
Hitia
Ebini
Malali
Ituni
Paradise
Kalkuni
Kwakwani
Orealla
Siparuta
Apoera
Epira
Matapi
Avanavero
Bakhuis
Raleigh Vallen Nature Preserve
Kapoesanti
Wanhatti
Akoelitatajeprati
Kabalebo
Suriname
Bakrakondre
Miles 0 50
Km 0 50
Map 28
Map data ©2024 Google
©2024 F. Espenak

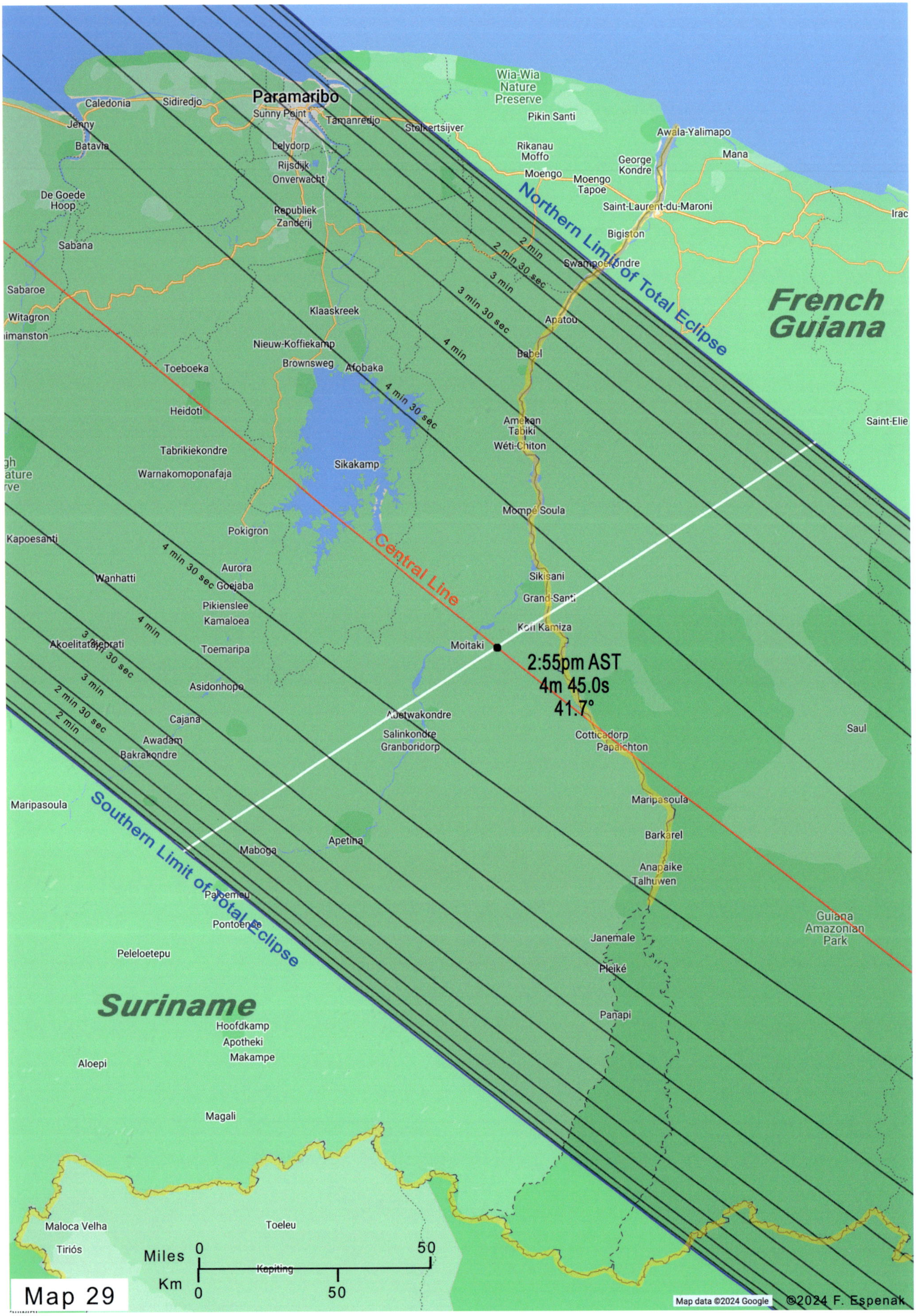
Paramaribo
Sunny Point
Tamanredjo
Caledonia
Sidiredjo
Jenny
Batavia
Lelydorp
Rijsdijk
Onverwacht
Republiek
Zanderij
De Goede
Hoop
Sabana
Sabaroe
Witagron
Klaaskreek
Nieuw-Koffiekamp
Brownsweg
Afobaka
Toeboeka
Heidoti
Tabrikiekondre
Warnakomoponafaja
Sikakamp
Pokigron
Kapoesanti
Aurora
Goejaba
Wanhatti
Pikienslee
Kamaloea
Toemaripa
Asidonhopo
Cajana
Awadam
Bakrakondre
Maripasoula
Maboga
Apetina
Paloemeu
Pontoenoe
Peleloetepu
Suriname
Hoofdkamp
Apotheki
Makampe
Aloepi
Magali
Maloca Velha
Tiriós
Toeleu
Kapiting
Wia-Wia
Nature
Preserve
Pikin Santi
Stolkertsijver
Rikanau
Moffo
Moengo
Moengo
Tapoe
George
Kondre
Awala-Yalimapo
Mana
Saint-Laurent-du-Maroni
Bigiston
Swampoe Kondre
Apatou
Babel
Amakan
Tabiki
Wéti-Chiton
Mompé Soula
Sikisani
Grand-Santi
Koti Kamiza
Moitaki
Saint-Elie
Saul
Cotticadorp
Papaichton
Maripasoula
Barkarel
Anapaike
Talhuwen
Janemale
Pleiké
Panapi
Guiana
Amazonian
Park
French
Guiana
Northern Limit of Total Eclipse
Southern Limit of Total Eclipse
Central Line
2 min
2 min 30 sec
3 min
3 min 30 sec
4 min
4 min 30 sec
2:55pm AST
4m 45.0s
41.7°
Miles 0 50
Km 0 50
Map 29
Map data ©2024 Google
©2024 F. Espenak

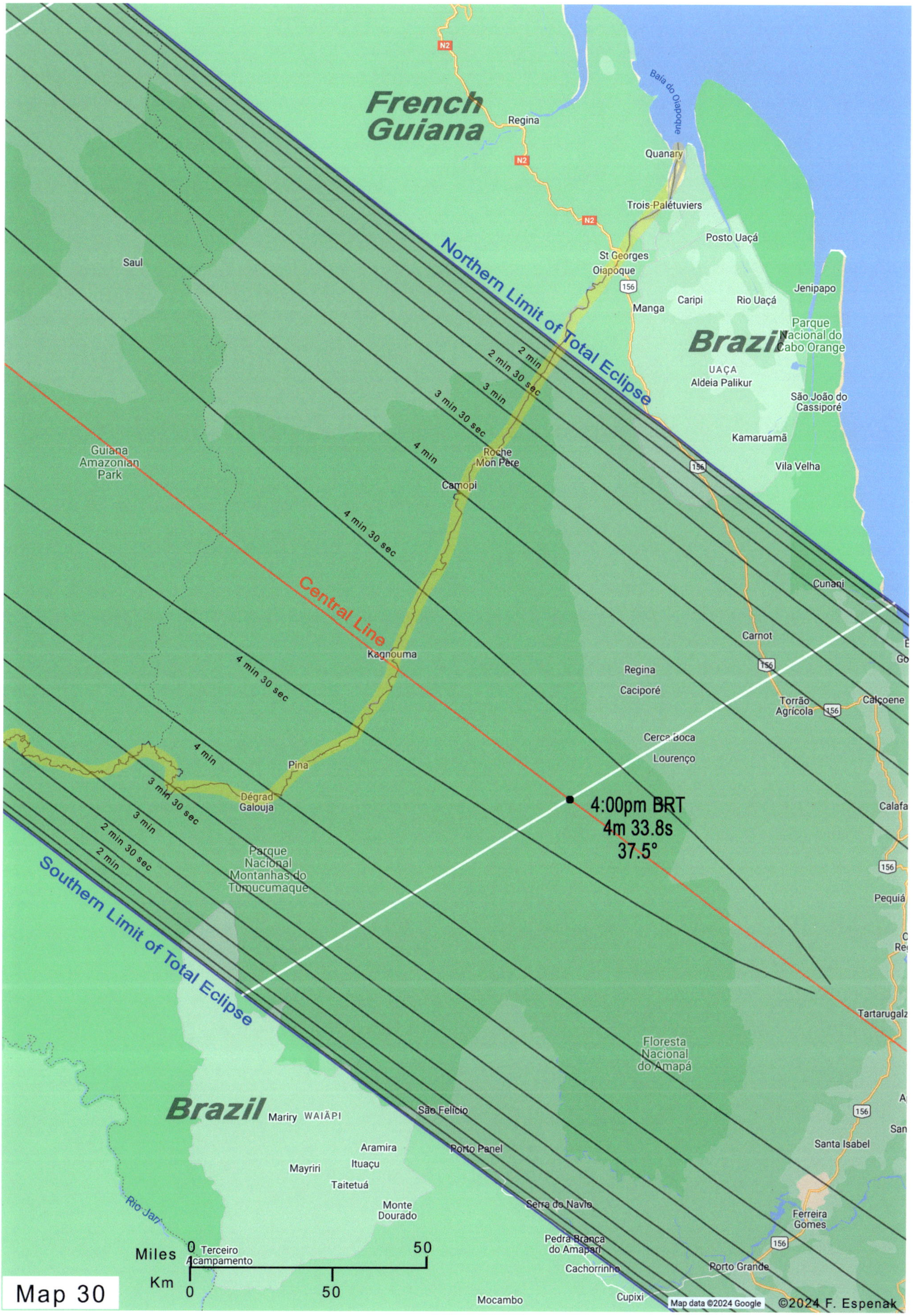
French Guiana
Brazil
Brazil
Northern Limit of Total Eclipse
Southern Limit of Total Eclipse
Central Line
2 min
2 min 30 sec
3 min
3 min 30 sec
4 min
4 min 30 sec
4 min 30 sec
4 min
3 min 30 sec
3 min
2 min 30 sec
2 min
4:00pm BRT
4m 33.8s
37.5°
Baía do Oiapoque
Regina
Quanary
Trois-Palétuviers
Posto Uaçá
St Georges
Oiapoque
Saul
Manga
Caripi
Rio Uaçá
Jenipapo
Parque Nacional do Cabo Orange
UAÇA
Aldeia Palikur
São João do Cassiporé
Kamaruamã
Roche Mon Pere
Vila Velha
Camopi
Guiana Amazonian Park
Cunani
Carnot
Kagnouma
Regina
Caciporé
Torrão Agrícola
Calçoene
Cerca Boca
Lourenço
Pina
Dégrad Galouja
Parque Nacional Montanhas do Tumucumaque
Pequiá
Tartarugalz
Floresta Nacional do Amapá
Mariry
WAIÃPI
São Felício
Santa Isabel
Aramira
Porto Panel
Mayriri
Ituaçu
Taitetuá
Rio Jari
Monte Dourado
Serra do Navio
Ferreira Gomes
Pedra Branca do Amapari
Terceiro Acampamento
Cachorrinho
Porto Grande
Mocambo
Cupixi
Miles 0 50
Km 0 50
Map 30
Map data ©2024 Google
©2024 F. Espenak

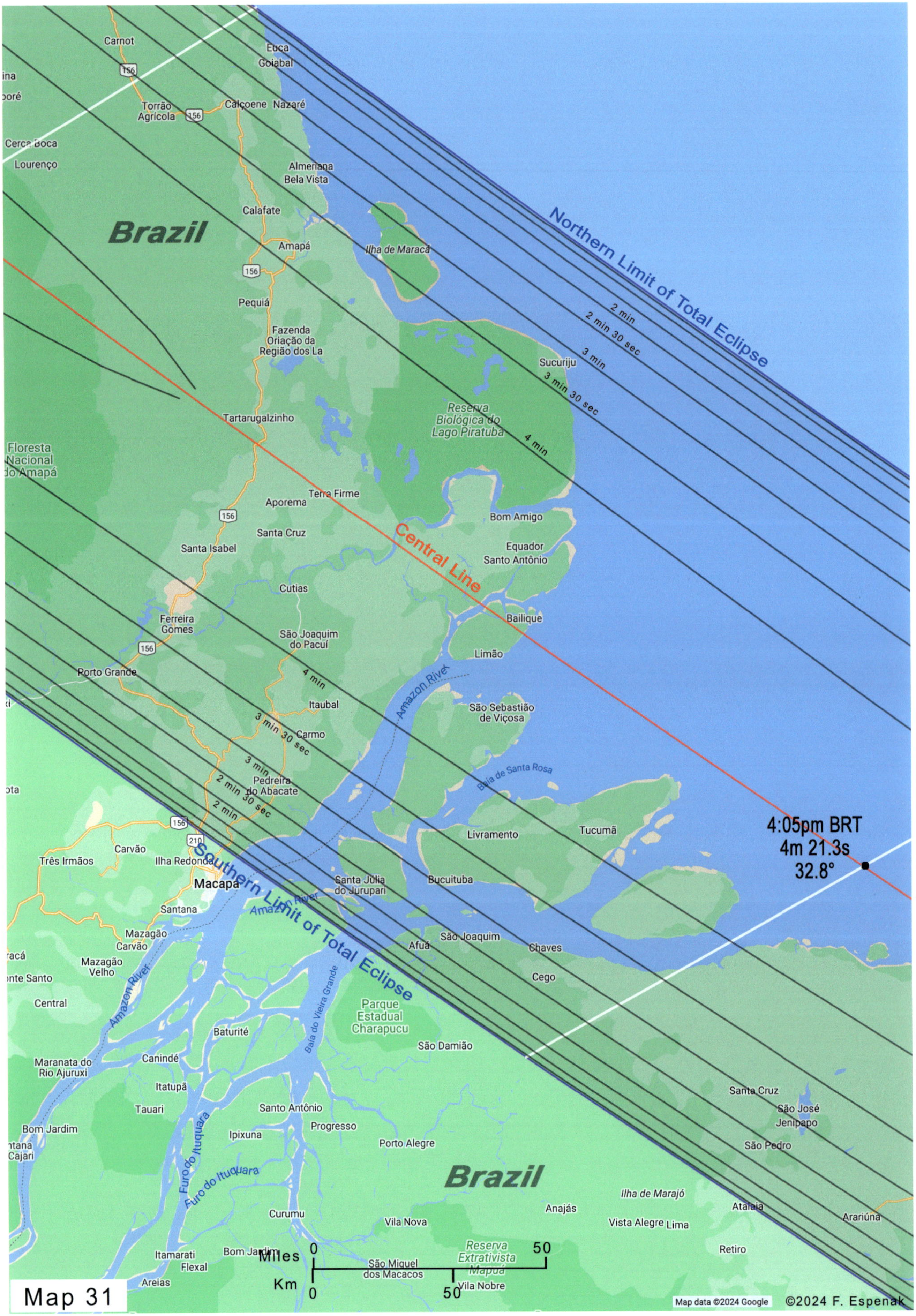
Brazil
Northern Limit of Total Eclipse
2 min
2 min 30 sec
3 min
3 min 30 sec
4 min
Central Line
Southern Limit of Total Eclipse
4:05pm BRT
4m 21.3s
32.8°
Carnot
Euca
Goiabal
Calçoene
Nazaré
Torrão Agrícola
Cerca Boca
Lourenço
Almeriana
Bela Vista
Calafate
Amapá
Ilha de Maracá
Pequiá
Fazenda Oriação da Região dos La
Sucuriju
Reserva Biológica do Lago Piratuba
Tartarugalzinho
Floresta Nacional do Amapá
Terra Firme
Aporema
Bom Amigo
Santa Cruz
Santa Isabel
Equador
Santo Antônio
Cutias
Bailique
Ferreira Gomes
São Joaquim do Pacuí
Limão
Porto Grande
Itaubal
Amazon River
São Sebastião de Viçosa
Carmo
Baía de Santa Rosa
Pedreira do Abacate
Livramento
Tucumã
Carvão
Três Irmãos
Ilha Redonda
Macapá
Santa Júlia do Jurupari
Bucuituba
Santana
Mazagão Carvão
Mazagão Velho
Afuá
São Joaquim
Chaves
Cego
Central
Parque Estadual Charapucu
Baía do Vieira Grande
Baturité
São Damião
Maranata do Rio Ajuruxi
Canindé
Itatupã
Tauari
Santo Antônio
Progresso
Bom Jardim
Ipixuna
Porto Alegre
Furo do Ituquara
Santa Cruz
São José
Jenipapo
São Pedro
Ilha de Marajó
Anajás
Atalaia
Curumu
Vila Nova
Vista Alegre
Lima
Ararиúna
Itamarati
Flexal
Bom Jardim
Retiro
Reserva Extrativista Mapuá
São Miguel dos Macacos
Vila Nobre
Areias
Miles 0 50
Km 0 50
Map 31
Map data ©2024 Google
©2024 F. Espenak

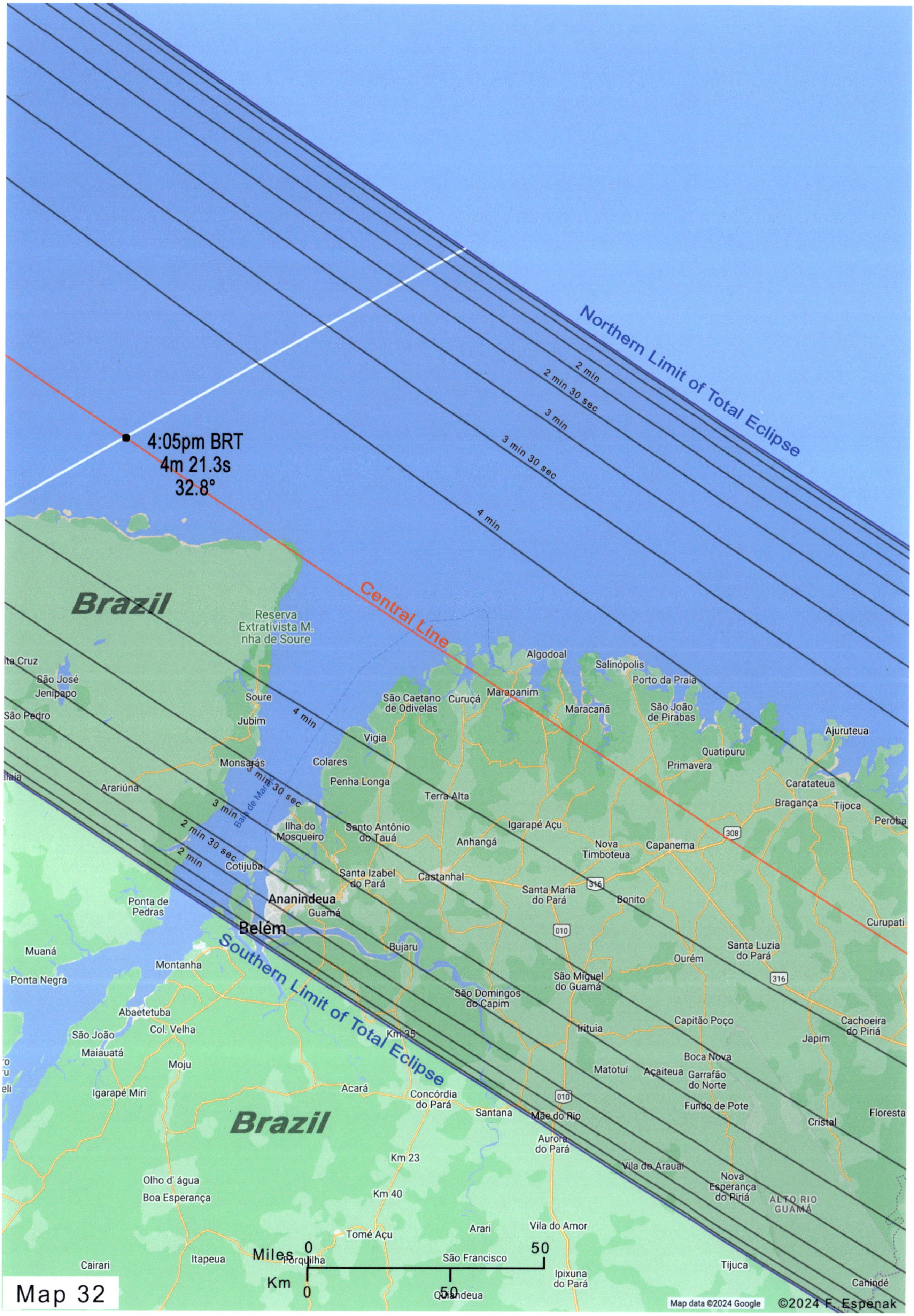
Northern Limit of Total Eclipse
2 min
2 min 30 sec
3 min
3 min 30 sec
4 min
4:05pm BRT
4m 21.3s
32.8°
Central Line
Brazil
Reserva Extrativista M. nha de Soure
4 min
3 min 30 sec
3 min
2 min 30 sec
2 min
Southern Limit of Total Eclipse
Brazil
Algodoal
Salinópolis
Porto da Praia
São Caetano de Odivelas
Curuçá
Marapanim
Maracanã
São João de Pirabas
Ajuruteua
Soure
Jubim
São José
Jenipapo
São Pedro
Vigia
Quatipuru
Primavera
Monsarás
Colares
Penha Longa
Caratateua
Terra Alta
Bragança
Tijoca
Peroba
Arariúna
Ilha do Mosqueiro
Santo Antônio do Tauá
Igarapé Açu
Anhangá
Nova Timboteua
Capanema
Cotijuba
Santa Izabel do Pará
Castanhal
Santa Maria do Pará
Bonito
Ponta de Pedras
Ananindeua
Guamá
Belém
Curupati
Muaná
Bujaru
Santa Luzia do Pará
Ourém
Montanha
Ponta Negra
São Miguel do Guamá
São Domingos do Capim
Abaetetuba
Col. Velha
Capitão Poço
Cachoeira do Piriá
São João
Maiauatá
Irituia
Km 35
Japim
Boca Nova
Moju
Matotuí
Açaiteua
Garrafão do Norte
Igarapé Miri
Acará
Concórdia do Pará
Fundo de Pote
Santana
Mãe do Rio
Cristal
Floresta
Aurora do Pará
Km 23
Vila do Arauaí
Olho d' água
Boa Esperança
Nova Esperança do Piriá
ALTO RIO GUAMÁ
Km 40
Arari
Vila do Amor
Tomé Açu
Cairari
Itapeua
Porquilha
São Francisco
Ipixuna do Pará
Tijuca
Quiandeua
Canindé
Miles 0 50
Km 0 50
Map 32
Map data ©2024 Google
©2024 F. Espenak

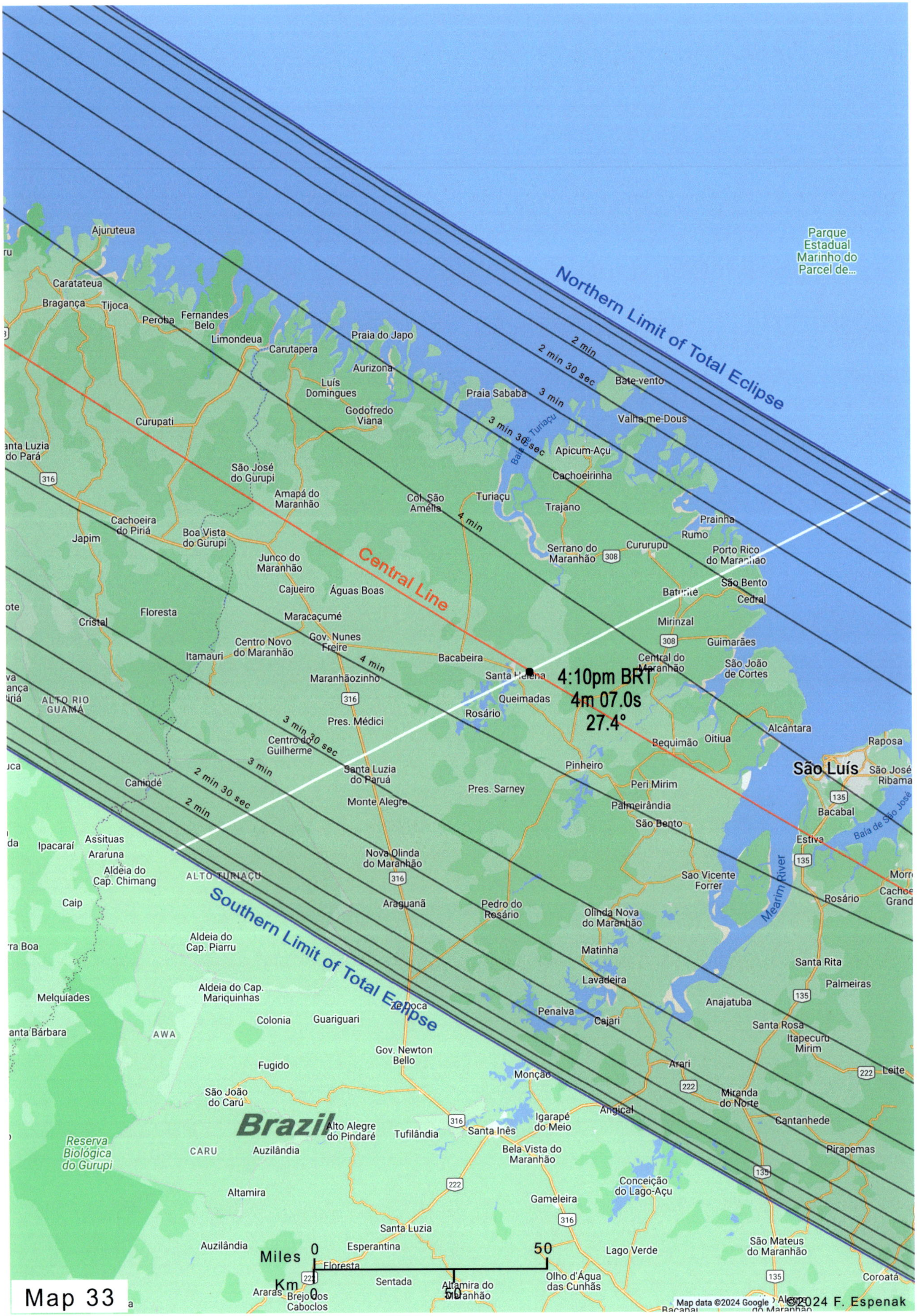
Northern Limit of Total Eclipse
Southern Limit of Total Eclipse
Central Line
2 min
2 min 30 sec
3 min
3 min 30 sec
4 min
4 min
3 min 30 sec
3 min
2 min 30 sec
2 min
4:10pm BRT
4m 07.0s
27.4°
Parque Estadual Marinho do Parcel de...
Ajuruteua
Caratateua
Bragança
Tijoca
Peroba
Fernandes Belo
Limondeua
Carutapera
Praia do Japo
Aurizona
Luís Domingues
Godofredo Viana
Praia Sababa
Bate-vento
Valha-me-Deus
Apicum-Açu
Cachoeirinha
Trajano
Curupati
São José do Gurupi
Amapá do Maranhão
Col. São Amélia
Turiaçu
Cachoeira do Piriá
Japim
Boa Vista do Gurupi
Junco do Maranhão
Serrano do Maranhão
Cururupu
Prainha
Rumo
Porto Rico do Maranhão
São Bento
Cedral
Cajueiro
Águas Boas
Floresta
Cristal
Maracaçumé
Mirinzal
Guimarães
Centro Novo do Maranhão
Gov. Nunes Freire
Itamauri
Bacabeira
Central do Maranhão
São João de Cortes
Maranhãozinho
Santa Helena
Queimadas
Rosário
Pres. Médici
Centro do Guilherme
Bequimão
Oitiua
Alcântara
Raposa
São Luís
Santa Luzia do Paruá
Pinheiro
Peri Mirim
Pres. Sarney
Monte Alegre
Palmeirândia
São Bento
Bacabal
Estiva
Baía de São José
Canindé
Assituas
Araruna
Ipacaraí
Aldeia do Cap. Chimang
ALTO RIO GUAMÁ
ALTO TURIAÇU
Nova Olinda do Maranhão
Sao Vicente Ferrer
Mearim River
Rosário
Caip
Araguanã
Pedro do Rosário
Olinda Nova do Maranhão
Aldeia do Cap. Piarru
Matinha
Santa Rita
Lavadeira
Palmeiras
Melquíades
Aldeia do Cap. Mariquinhas
Anajatuba
Colonia
Guariguari
Zé Doca
Penalva
Cajari
Santa Rosa
Itapecuru Mirim
AWA
Gov. Newton Bello
Fugido
Monção
Arari
Leite
São João do Carú
Miranda do Norte
Angical
Brazil
Alto Alegre do Pindaré
Tufilândia
Santa Inês
Igarapé do Meio
Cantanhede
Reserva Biológica do Gurupi
CARU
Auzilândia
Bela Vista do Maranhão
Pirapemas
Altamira
Conceição do Lago-Açu
Gameleira
Santa Luzia
Auzilândia
Esperantina
Lago Verde
São Mateus do Maranhão
Miles 0 50
Km 0 50
Floresta
Sentada
Altamira do Maranhão
Olho d'Água das Cunhãs
Coroatá
Araras
Brejo dos Caboclos
Map 33
Map data ©2024 Google
©2024 F. Espenak

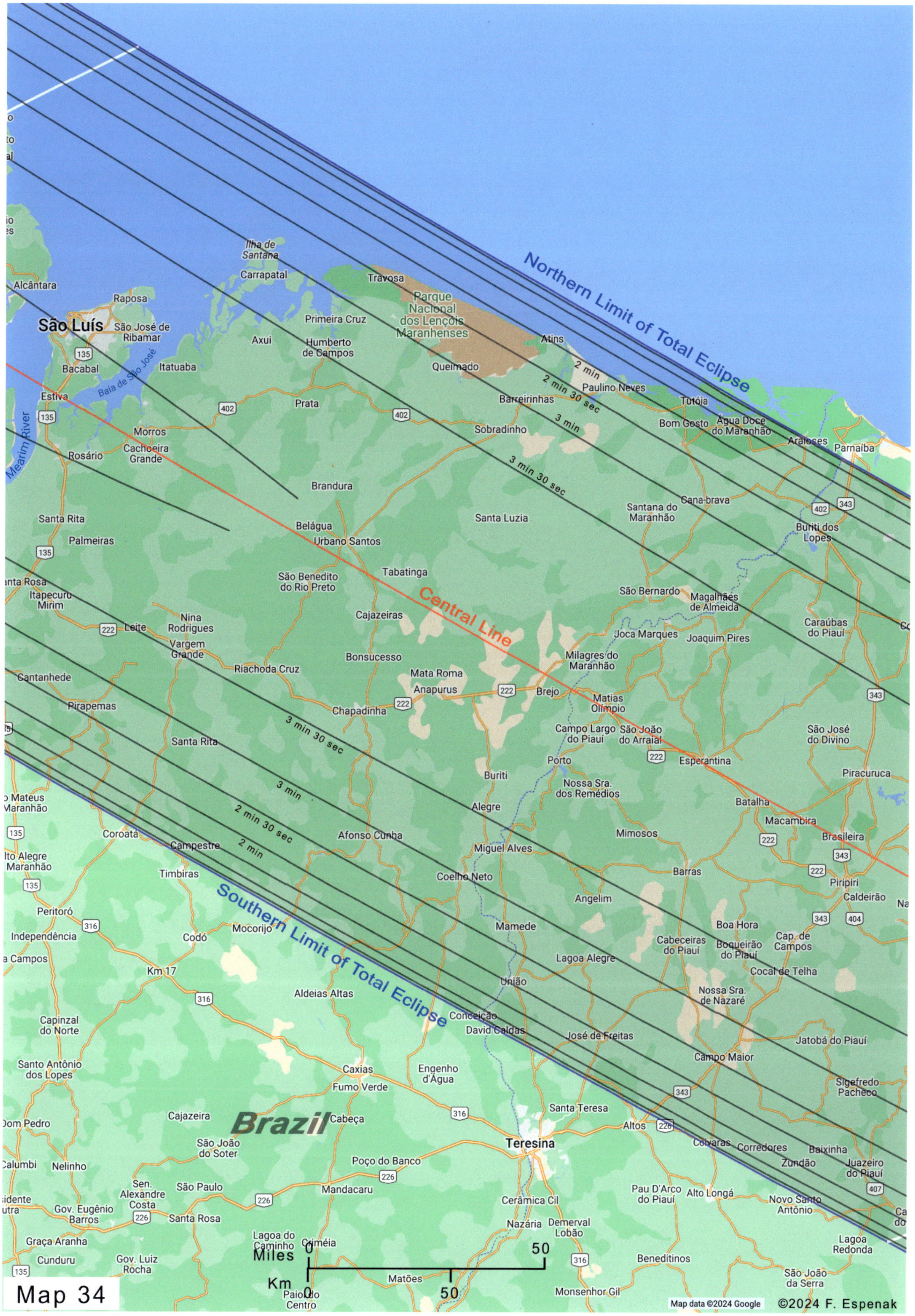
Northern Limit of Total Eclipse
Central Line
Southern Limit of Total Eclipse
2 min
2 min 30 sec
3 min
3 min 30 sec
Brazil
São Luís
Alcântara
Raposa
São José de Ribamar
Bacabal
Estiva
Baia de São José
Itatuaba
Mearim River
Morros
Cachoeira Grande
Rosário
Ilha de Santana
Carrapatal
Primeira Cruz
Axui
Humberto de Campos
Travosa
Parque Nacional dos Lençóis Maranhenses
Queimado
Atins
Paulino Neves
Barreirinhas
Prata
Sobradinho
Tutóia
Bom Gosto
Água Doce do Maranhão
Araioses
Parnaíba
Brandura
Santa Rita
Palmeiras
Belágua
Urbano Santos
Santa Luzia
Santana do Maranhão
Cana-brava
Buriti dos Lopes
Tabatinga
São Benedito do Rio Preto
São Bernardo
Magalhães de Almeida
Itapecuru Mirim
Nina Rodrigues
Leite
Vargem Grande
Cajazeiras
Joca Marques
Joaquim Pires
Caraúbas do Piauí
Bonsucesso
Riachoda Cruz
Milagres do Maranhão
Mata Roma
Anapurus
Brejo
Matias Olímpio
Cantanhede
Pirapemas
Chapadinha
Campo Largo do Piauí
São João do Arraial
São José do Divino
Santa Rita
Porto
Esperantina
Buriti
Nossa Sra. dos Remédios
Piracuruca
Batalha
Macambira
Alegre
Mimosos
Brasileira
Coroatá
Campestre
Afonso Cunha
Miguel Alves
Barras
Piripiri
Timbiras
Coelho Neto
Angelim
Caldeirão
Peritoró
Mamede
Boa Hora
Independência
Codó
Mocorijo
Cabeceiras do Piauí
Boqueirão do Piauí
Cap. de Campos
Lagoa Alegre
Cocal de Telha
Km 17
União
Aldeias Altas
Nossa Sra. de Nazaré
Capinzal do Norte
Conceição
David Caldas
José de Freitas
Jatobá do Piauí
Campo Maior
Santo Antônio dos Lopes
Caxias
Fumo Verde
Engenho d'Água
Sigefredo Pacheco
Cajazeira
Cabeça
Santa Teresa
Altos
Teresina
Colvaras
Corredores
Baixinha
São João do Soter
Poço do Banco
Zundão
Juazeiro do Piauí
Nelinho
Sen. Alexandre Costa
São Paulo
Mandacaru
Cerâmica Cil
Pau D'Arco do Piauí
Alto Longá
Novo Santo Antônio
Gov. Eugênio Barros
Santa Rosa
Nazária
Demerval Lobão
Graça Aranha
Lagoa do Caminho
Criméia
Lagoa Redonda
Cunduru
Gov. Luiz Rocha
Beneditinos
São João da Serra
Matões
Paiol do Centro
Monsenhor Gil
Miles 0 50
Km 0 50
Map 34
Map data ©2024 Google
©2024 F. Espenak

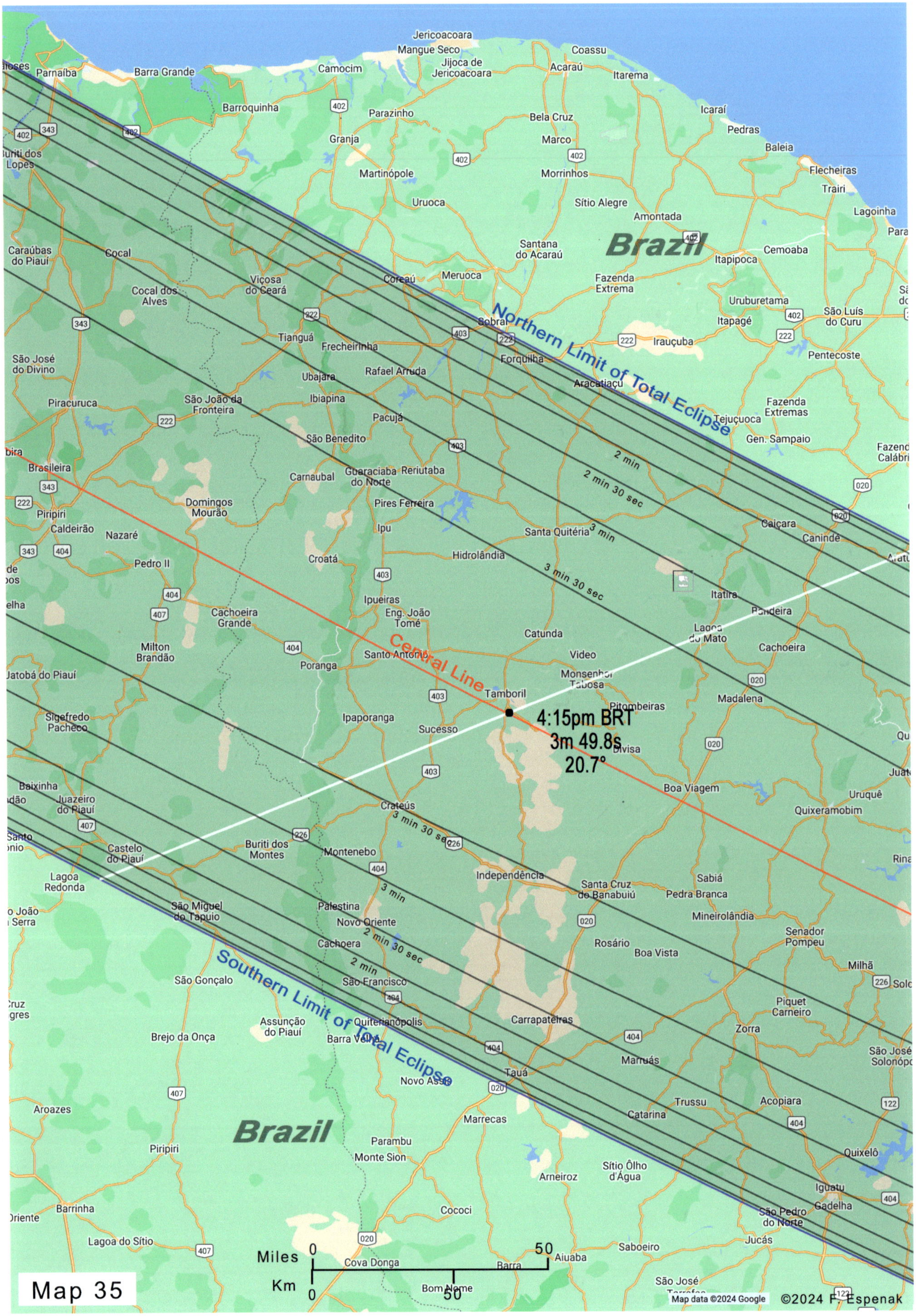
Brazil
Northern Limit of Total Eclipse
Central Line
Southern Limit of Total Eclipse
2 min
2 min 30 sec
3 min
3 min 30 sec
4:15pm BRT
3m 49.8s
20.7°
Jericoacoara
Mangue Seco
Jijoca de Jericoacoara
Camocim
Barra Grande
Parnaíba
Coassu
Acaraú
Itarema
Barroquinha
Parazinho
Bela Cruz
Marco
Granja
Icaraí
Pedras
Baleia
Martinópole
Morrinhos
Flecheiras
Trairi
Uruoca
Sítio Alegre
Amontada
Lagoinha
Buriti dos Lopes
Caraúbas do Piauí
Cocal
Cocal dos Alves
Viçosa do Ceará
Coreaú
Meruoca
Santana do Acaraú
Fazenda Extrema
Itapipoca
Cemoaba
Uruburetama
Itapagé
São Luís do Curu
Sobral
Irauçuba
Pentecoste
Tianguá
Frecheirinha
Forquilha
São José do Divino
Ubajara
Rafael Arruda
Aracatiaçu
Ibiapina
Piracuruca
São João da Fronteira
Pacujá
Fazenda Extremas
Tejuçuoca
Gen. Sampaio
São Benedito
Brasileira
Carnaubal
Guaraciaba do Norte
Reriutaba
Piripiri
Domingos Mourão
Pires Ferreira
Caldeirão
Nazaré
Ipu
Santa Quitéria
Caiçara
Canindé
Croatá
Hidrolândia
Pedro II
Itatira
Ipueiras
Eng. João Tomé
Bandeira
Cachoeira Grande
Lagoa do Mato
Catunda
Cachoeira
Milton Brandão
Poranga
Santo Antônio
Video
Monsenhor Tabosa
Jatobá do Piauí
Tamboril
Madalena
Pitombeiras
Sigefredo Pacheco
Ipaporanga
Sucesso
Divisa
Baixinha
Boa Viagem
Uruquê
Juazeiro do Piauí
Crateús
Quixeramobim
Castelo do Piauí
Buriti dos Montes
Montenebo
Lagoa Redonda
Independência
Santa Cruz do Banabuiú
Sabiá
Pedra Branca
São Miguel do Tapuio
Palestina
Mineirolândia
Novo Oriente
Senador Pompeu
Cachoeira
Rosário
Boa Vista
São Gonçalo
São Francisco
Milhã
Piquet Carneiro
Assunção do Piauí
Quiterianópolis
Carrapateiras
Brejo da Onça
Barra Velha
Zorra
Marruás
São José Solonópole
Tauá
Novo Assis
Aroazes
Trussu
Acopiara
Catarina
Marrecas
Parambu
Monte Sion
Quixelô
Piripiri
Sítio Ôlho d'Água
Arneiroz
Iguatu
Gadelha
Barrinha
São Pedro do Norte
Cococi
Lagoa do Sítio
Saboeiro
Jucás
Cova Donga
Barra
Aiuaba
Bom Nome
São José
Miles 0 50
Km 0 50
Map 35
Map data ©2024 Google
©2024 F. Espenak

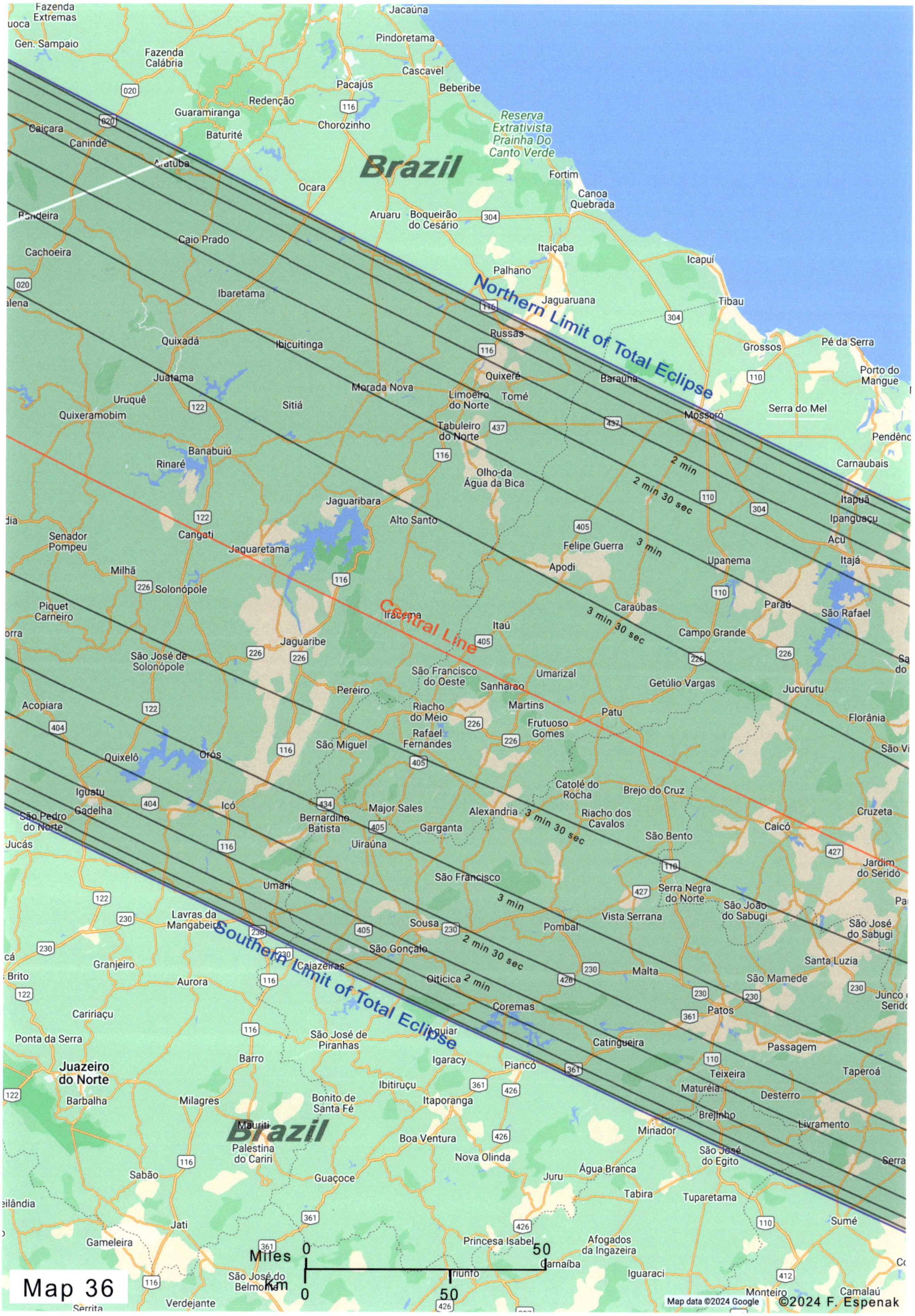
Brazil
Northern Limit of Total Eclipse
Central Line
Southern Limit of Total Eclipse
2 min
2 min 30 sec
3 min
3 min 30 sec
3 min 30 sec
3 min
2 min 30 sec
2 min
Fazenda Extremas
Gen. Sampaio
Fazenda Calábria
Jacaúna
Pindoretama
Cascavel
Pacajús
Beberibe
Redenção
Guaramiranga
Baturité
Chorozinho
Caiçara
Canindé
Aratuba
Reserva Extrativista Prainha Do Canto Verde
Fortim
Ocara
Canoa Quebrada
Aruaru
Boqueirão do Cesário
Caio Prado
Cachoeira
Itaiçaba
Palhano
Icapuí
Ibaretama
Jaguaruana
Tibau
Russas
Grossos
Pé da Serra
Quixadá
Ibicuitinga
Quixeré
Baraúna
Porto do Mangue
Juatama
Morada Nova
Limoeiro do Norte
Tomé
Uruquê
Sitiá
Serra do Mel
Quixeramobim
Mossoró
Tabuleiro do Norte
Pendências
Banabuiú
Rinaré
Olho-da Água da Bica
Carnaubais
Jaguaribara
Itapuã
Senador Pompeu
Cangati
Alto Santo
Ipanguaçu
Jaguaretama
Felipe Guerra
Açu
Milhã
Apodi
Upanema
Itajá
Solonópole
Piquet Carneiro
Iracema
Caraúbas
Paraú
São Rafael
Jaguaribe
Itaú
Campo Grande
São José de Solonópole
São Francisco do Oeste
Umarizal
Getúlio Vargas
Pereiro
Sanharão
Jucurutu
Acopiara
Riacho do Meio
Martins
Frutuoso Gomes
Patu
Florânia
Rafael Fernandes
São Miguel
Quixelô
Orós
Catolé do Rocha
Brejo do Cruz
Iguatu
Gadelha
Icó
Bernardino Batista
Major Sales
Alexandria
Riacho dos Cavalos
Cruzeta
São Pedro do Norte
Garganta
São Bento
Caicó
Jucás
Uiraúna
Jardim do Seridó
São Francisco
Serra Negra do Norte
Umari
São João do Sabugi
Vista Serrana
Lavras da Mangabeira
Sousa
Pombal
São José do Sabugi
São Gonçalo
Granjeiro
Cajazeiras
Malta
Santa Luzia
Aurora
Oiticica
São Mamede
Patos
Cariríaçu
Coremas
Ponta da Serra
São José de Piranhas
Aguiar
Catingueira
Passagem
Juazeiro do Norte
Barro
Igaracy
Piancó
Teixeira
Taperoá
Barbalha
Milagres
Ibitiruçu
Bonito de Santa Fé
Itaporanga
Maturéia
Desterro
Brejinho
Livramento
Mauriti
Palestina do Cariri
Boa Ventura
Minador
São José do Egito
Nova Olinda
Sabão
Guaçoce
Juru
Água Branca
Tabira
Tuparetama
Jati
Sumé
Gameleira
Princesa Isabel
Afogados da Ingazeira
Carnaíba
Iguaraci
Triunfo
São José do Belmonte
Verdejante
Monteiro
Camalaú
Miles
0
50
Km
0
50
Map 36
Map data ©2024 Google
©2024 F. Espenak

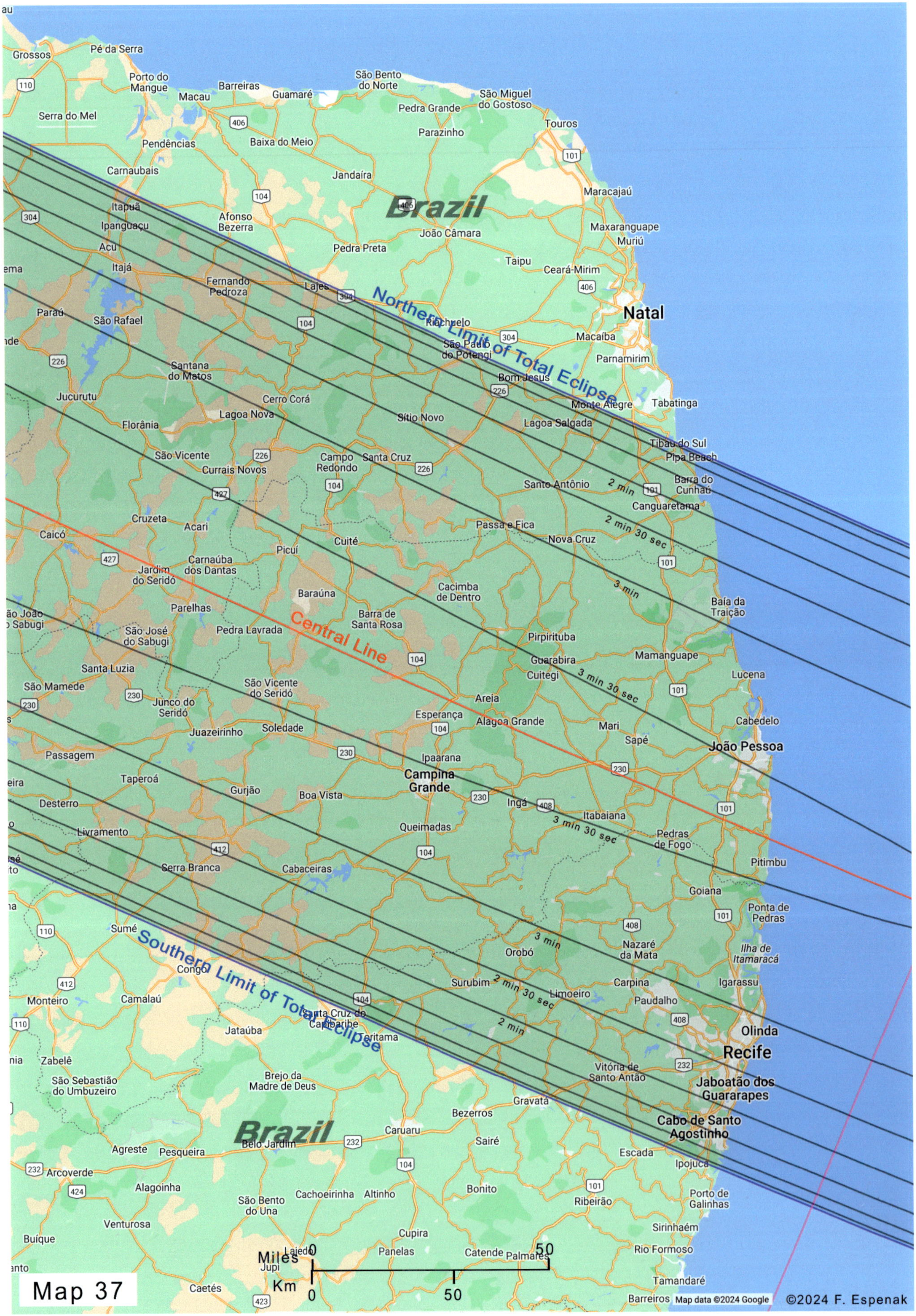
Brazil
Northern Limit of Total Eclipse
Central Line
Southern Limit of Total Eclipse
2 min
2 min 30 sec
3 min
3 min 30 sec
3 min 30 sec
3 min
2 min 30 sec
2 min
Natal
João Pessoa
Campina Grande
Olinda
Recife
Jaboatão dos Guararapes
Cabo de Santo Agostinho
Brazil
Miles 0 50
Km 0 50
Map 37
Map data ©2024 Google
©2024 F. Espenak

Eclipse Guides from AstroPixels Publishing

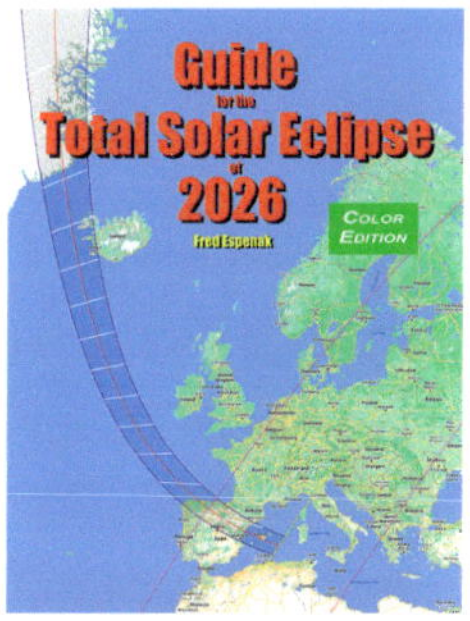

Guide for the Total Solar Eclipse of 2026

The total solar eclipse of August 12, 2026 will be visible from a path across northern Russia, the Arctic, Greenland, Iceland, and Spain. The **Guide for the Total Solar Eclipse of 2026** contains 14 full-page maps of the path. The large scale shows roads, towns and cities, rivers and lakes.

Tables for hundreds of cities provide times each eclipse phase of the along with the eclipse magnitude, duration, and the Sun's altitude. A climatology study identifies areas along the eclipse path where the highest probability of favorable weather may be found.

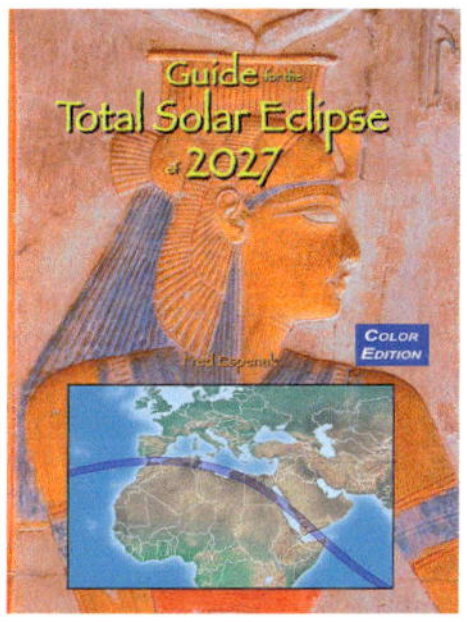

Guide for the Total Solar Eclipse of 2027

The total solar eclipse of August 2, 2027 will be visible from a path across Spain, Morocco, Algeria, Tunisia, Libya, Egypt, Sudan, Saudia Arabia, Yemen, and Somalia. The **Guide for the Total Solar Eclipse of 2027** contains 20 full-page maps of the path.

Tables for hundreds of cities provide times for each eclipse phase along with the eclipse magnitude, duration and the Sun's altitude. Cloud statistics identify areas where the highest probability of clear sky may be found.

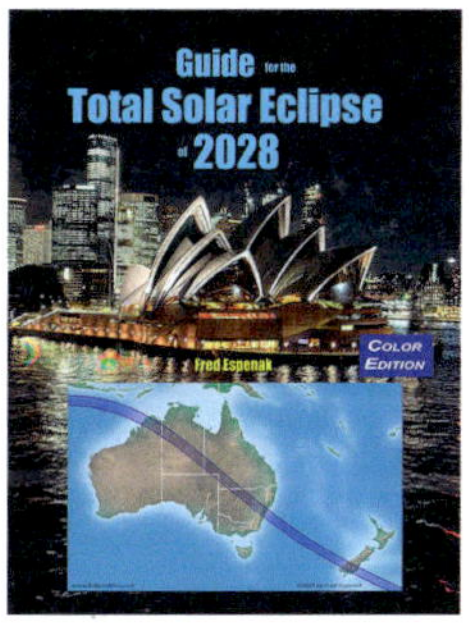

Guide for the Total Solar Eclipse of 2028

The total solar eclipse of July 22, 2028 will be visible from a path across Australia and New Zealand. The **Guide for the Total Solar Eclipse of 2028** contains 19 full-page maps of the path of totality.

Tables for over one hundred cities provide times for each eclipse phase with the duration of totality and the Sun's altitude. Cloud statistics identify areas where the highest probability of clear sky may be found. A preview is given for total eclipses through 2039.

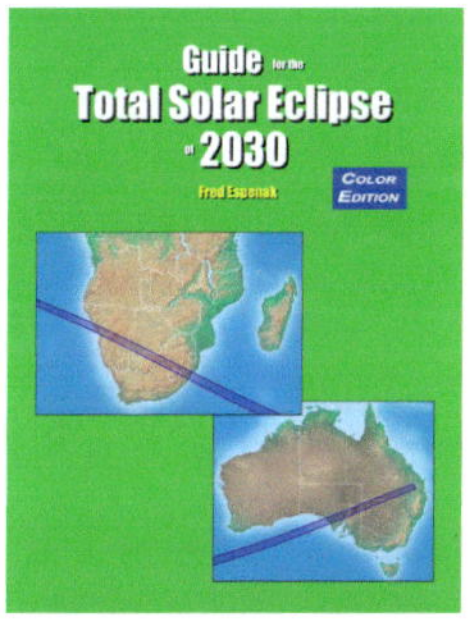

Guide for the Total Solar Eclipse of 2030

The total solar eclipse of November 25, 2030 will be visible from a path across Africa and Australia. The **Guide for the Total Solar Eclipse of 2028** contains 17 full-page maps of the path of totality.

Tables for over one hundred cities provide times for each eclipse phase with the duration of totality and the Sun's altitude. Cloud statistics identify areas where the highest probability of clear sky may be found. A preview is given for total eclipses through 2039.

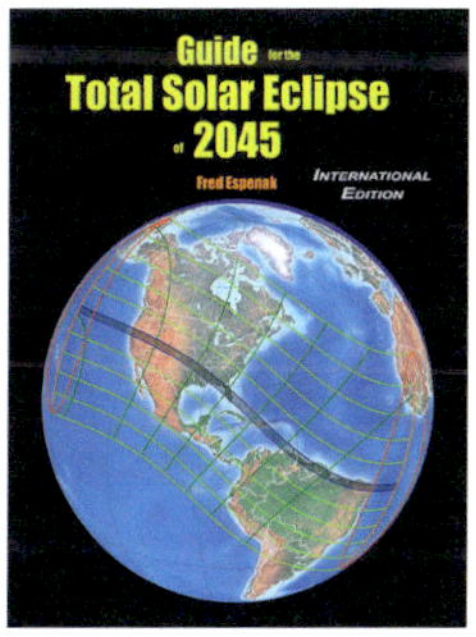

Guide for the Total Solar Eclipse of 2045

The total solar eclipse of August 12, 2045 will be visible from a path across the USA, the Caribbean, and South America. The **Guide for the Total Solar Eclipse of 2045** contains 37 full-page maps of the path of totality.

Tables for hundreds cities provide times for each eclipse phase with the duration of totality, eclipse magnitude, and the Sun's altitude. Cloud statistics identify areas where the highest probability of clear sky may be found.

For more information on each *Guide*, including sample pages of each, visit:

astropixels.com/pubs/guides.html

The *Guides* may be ordered from ***Amazon***

Eclipse Books from AstroPixels Publishing

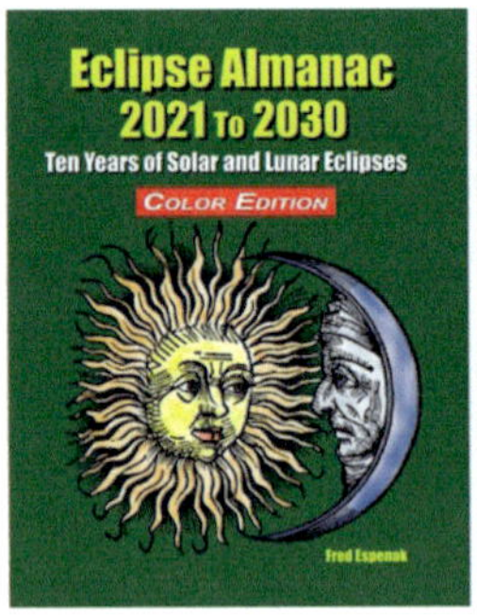

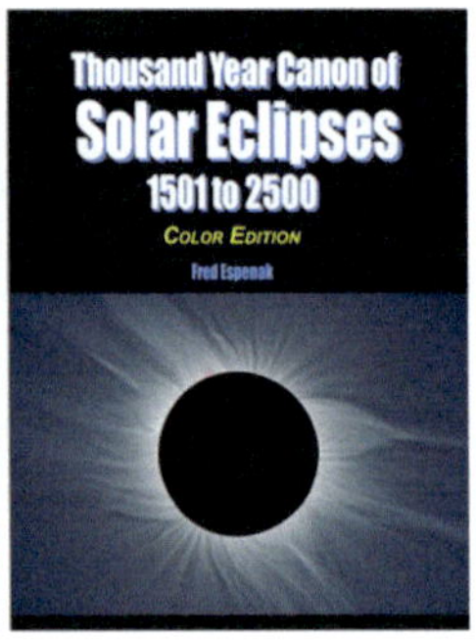

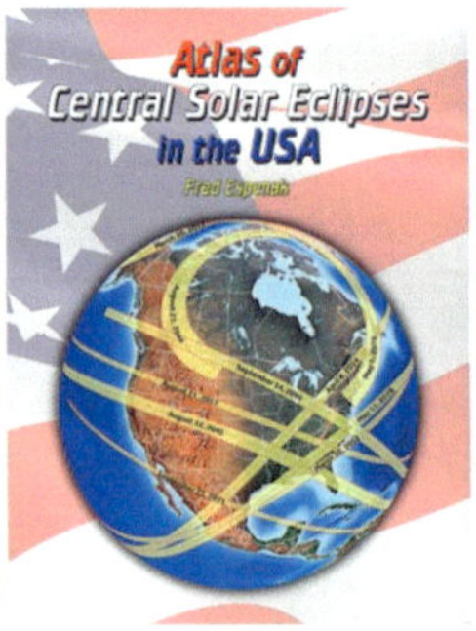

Eclipse Almanac – 2021 to 2030

A concise reference for every eclipse of the Sun and the Moon over a 10-year period. It identifies when and where each event will be seen with maps for each. Additional details about each eclipse are included. Other volumes in this series include the decades 2032 to 2040, 2041 to 2050, 2051 to 2060, and 2061 to 2070.

21st Century Canon of Solar Eclipses

The complete guide to every solar eclipse occurring from 2001 tom 2100 (224 eclipses in all). It includes information and maps for all total, annular, hybrid, and partial eclipses. A special world atlas shows detailed, full page maps of all central eclipse paths (total, annular, and hybrid).

Thousand Year Canon of Solar Eclipses — 1501 to 2500

Contains maps and data for each of the 2,389 solar eclipses occurring over the ten-century period centered on the present era. The maps are arranged twelve to a page at an image scale allowing the determination of eclipse visibility from any location on Earth. A comprehensive catalog lists the essential characteristics of each eclipse.

Atlas of Central Solar Eclipses in the USA

When was the last total eclipse through the USA and when is the next? How often do they happen? What total eclipse tracks passed across the USA during the 17th, 18th, and 19th centuries, etc., and what states did they include? And how often is a total solar eclipse visible from each of the 50 states? The Atlas of Central Solar Eclipses in the USA answers all these questions and more, with hundreds of maps and tables.

21st Century Canon of Lunar Eclipses

The complete guide to every lunar eclipse occurring from 2001 tom 2100 (228 eclipses in all). It includes information and maps for all total, partial, and penumbral eclipses. The predictions use a new model for Earth's elliptical shadows.

Thousand Year Canon of Lunar Eclipses — 1501 to 2500

Contains maps and data for each of the 2,424 lunar eclipses occurring over the ten-century period centered on the present era. The maps are arranged twelve to a page at an image scale allowing the determination of eclipse visibility from any location on Earth. A comprehensive catalog lists the essential characteristics of each eclipse.

All books are available in two editions: 1) Black and White, and 2) Color.

For more information including sample pages of each, visit:

astropixels.com/pubs/guides.html

Each *Book* may be ordered from ***Amazon***

Made in the USA
Monee, IL
23 August 2026

61038670R00038